T0325895

Designing SCADA Application Software

Designing SCADA Application Software

A Practical Approach

Stuart G. McCrady

Technical Instructor in SCADA Technology, ON, Canada

AMSTERDAM • BOSTON • HEIDELBERG • LONDON • NEW YORK • OXFORD
PARIS • SAN DIEGO • SAN FRANCISCO • SINGAPORE • SYDNEY • TOKYO

Elsevier
32 Jamestown Road, London NW1 7BY, UK
225 Wyman Street, Waltham, MA 02451, USA

First edition 2013

Notices

Knowledge and best practice in this field are constantly changing. As new research and experience broaden our understanding, changes in research methods, professional practices, or medical treatment may become necessary.

Practitioners and researchers must always rely on their own experience and knowledge in evaluating and using any information, methods, compounds, or experiments described herein. In using such information or methods they should be mindful of their own safety and the safety of others, including parties for whom they have a professional responsibility.

To the fullest extent of the law, neither the Publisher nor the authors, contributors, or editors, assume any liability for any injury and/or damage to persons or property as a matter of products liability, negligence or otherwise, or from any use or operation of any methods, products, instructions, or ideas contained in the material herein.

British Library Cataloguing-in-Publication Data
A catalogue record for this book is available from the British Library

Library of Congress Cataloging-in-Publication Data
A catalog record for this book is available from the Library of Congress

ISBN: 978-0-12-417000-1

For information on all Elsevier publications visit our website at store.elsevier.com

This book has been manufactured using Print On Demand technology. Each copy is produced to order and is limited to black ink. The online version of this book will show color figures where appropriate.

Contents

About the Author

Stuart McCrady is a Certified Engineering Technologist in the field of Electronics and Physics Engineering, as well as a Certified Professional Educator in the field of technical training for adults. He spent the first two years of his career in electronics, installing and servicing both large mainframe computer systems and small minicomputers. He then shifted to software programming in automation systems. This field of minicomputer programming required developing application software in machine or assembly language, executing at the hardware level. Field devices such as limit switches, pushbuttons, and solenoid valves, were connected to custom designed hardware interface boards installed inside the minicomputers. From the minicomputers of the 1970s to the PLCs and HMIs of today, Stuart has worked with a broad range of technologies using a variety of hardware and software platforms. He was involved in the design and implementation of more than 50 SCADA type projects. As his career progressed, Stuart acquired both more experience and more responsibility in the field of system integration and SCADA systems consulting. Stuart has served as programmer, project leader, project administrator, consultant, department manager, and SCADA system designer.

Throughout his career, Stuart strove to establish programming standards and design methodologies that could be applied to any SCADA application. In the mid-1970s, he developed a program design and documentation system which he called FLOCODE, which resembled high level languages such as C, but was written in plain English. The purpose of the system was to allow the programmer to design software using English-like statements using structured programming constructs such as: If-Then-Else and Do-While/Do-Until. He applied this method to his own programming at both the high level language and the machine level language; this design documentation then became the comments and program description once the program was completed.

Later, Stuart was involved in the establishment of a systematic tagging system for signal names which would work both for hardware signals as well as internal software points. In addition, a system of structured descriptions was developed for the PLCs which described the operations in simple English, but referenced key signals and operating parameters; this documentation served as the design document for the PLC programming. Stuart expanded this combination of systems into the complete design and documentation system which is the focus of his book: 'Designing SCADA Application Software: A Practical Approach'.

In addition to this book on Designing SCADA Application Software, Stuart has published articles in trade magazines, as well as presented a paper on the application of computer control systems in the water treatment plants at an American Water Works Association convention.

In 2006, Stuart made another shift in his career, becoming a full time instructor/trainer. Stuart taught courses at the community college level and at the trade school level; courses included: electronics, residential wiring, digital logic circuits, communication networks, electro-pneumatic control systems, electrical motors and motor control circuits, and PLC programming. Since 2011, Stuart has been traveling throughout Canada and the United States, teaching PLC and HMI programming in cities across both countries. His extensive experience in the industry has served him well in the classrooms, as he is able to bring real world experiences into the classroom such that the students not only understand the programming material, but also understand how the concepts are applied.

Preface

Today's SCADA systems incorporate Programmable Logic Controllers (PLCs), Human–Machine Interface (HMI) workstations, and network communication systems into a complete integrated system. Each of the major components requires one or more form of programming from program logic to configuration to process graphic displays to communication configuration. While there are books on the market which teach PLC programming, there is a need for a comprehensive book that addresses the programming involved in all parts of a SCADA system.

Typically, the books available on PLC programming offer the reader detailed explanations of each of the instructions available for the particular brand of PLC. Short code sections are presented to illustrate how the instructions are used and how they function. Unfortunately, there is nothing in these books which explains to the reader how to organize and design the application software as a complete project, how to structure the database points, and how to document the program logic design. In short, there is a need for something that explains not only how SCADA application software should be developed, but also that explains how to create the type of software documentation required by today's customers.

Over a period of 35+ years, I have been involved in all aspects of SCADA systems: consulting, system design, programming, commissioning, training, and documentation. As a result of working on more than 50 such projects, a number of simple and efficient methods have evolved for designing and programming these systems. By following these methods, one can efficiently design, program, and implement the programming of all of the aspects of the SCADA system. A simple yet effective documentation system is used to ensure that not only is every part of the system designed properly, but also a thorough and complete documentation set is produced.

It is interesting that as I have been championing standards for software design and documentation, that more and more customers are establishing their own standards, all of which are based upon the ones that I have been using. It is rewarding to see many of the customers for whom I did SCADA programming have now created their own software standards, which are built upon those standards and concepts which I helped the customers to adopt. This is not to say that I single handedly caused these customers to follow my standards, but rather that I and others have established the need for good documentation, and that many of the standards which I used for these customers have now evolved into well-established standards. Anyone doing programming for these customers must adhere to their standards, and it is these standards that are presented in this book.

SCADA systems are global; every automated system in the world involves some form of process controller and user interface facility. These systems can range from a single controller that monitors and controls a small set of processes with a single workstation for user interfacing, to large geographical systems of controllers, user interface workstations, server computers, and both local and system wide communication networks. The user interface, in the form of animated process graphic displays, present current and historical operating information to the user, allowing users to observe system operations, as well as interact to effect control actions to the system. This user interface is functionally the same everywhere, regardless of the country; only the language would differ to suit the country.

The International Electrotechnical Commission, or IEC, has established a standard for programming of the field controllers, which has been adopted by most major controller manufacturers. This standard, IEC 61131-3, has defined five languages in which the controller programming can be done; a brief explanation of each language is provided in this book. In Canada and the United States, the controllers are typically programmed in 'Ladder Logic', which closely resembles electrical wiring diagrams. In Europe, a machine level language, 'Instruction List' or 'Machine Language', is the preferred language for field controllers.

Speaking personally, I have programmed with three of the five languages. The goal of this standard is to make the controller programming independent of the manufacturer; if one can program in Ladder Logic on one manufacturer's controller, then he or she can do so on another manufacturer's controller. When manufacturers adhere to these programming standards, then programmers can 'port' their software design across platforms; there will always be some changes and modifications required, but the essence of the program logic that was developed can be applied to any controller. One of the benefits of this standardization is that the system integrator does not have to learn a new language each time they switch to another manufacturer's product.

As an aside, the IEC is one of three key organizations which establish such standards; the other two are the International Telecommunication Union (ITU) and the International Organization for Standardization (ISO). An example of an ISO standards that is used quite commonly is the 7-layer model for communications; this standard has been adopted by most manufacturers for the interface driver communications. These three organizations work closely to ensure that any standard created works seamlessly across systems.

The methods and techniques presented in this book are based upon my own experience working in the Canadian market; but these methods apply to any system in any country. Every automation project needs software designed and developed for that particular application, and software documentation for this software is imperative. These methods of designing and developing application software are global, and can be applied to any project involving process controllers and user interface software systems, anywhere in the world. So regardless of location in the world, automation systems require proper software design with complete documentation.

This book is about how to design and develop application software for SCADA systems. Starting with the first chapter, the need for programming standards is established by explaining the longevity of SCADA systems, and therefore the need to develop a design which can be carried through the changing technologies. The remaining chapters then address the elements of SCADA software from the perspective of what is being designed and how it should be designed and developed.

Following is a brief summary of each of the chapters of this book, so as to provide the reader with an overview of the material in this book, and serve as an introduction to the material presented.

Chapter 1: Introduction

This chapter provides an explanation of SCADA software with respect to the intent of this book. This chapter highlights the importance of developing application software as part of the complete software development process. It also indicates that software documentation can be created 'after the fact' in a reverse engineering fashion, so that any system can be documented.

Chapter 2: The Elements of SCADA Software

This chapter identifies the various aspects of application software found in today's SCADA systems. As will be seen, a SCADA system involves different programming facilities at all of the levels of the system. The software is not just the field controller program listing, but includes database documentation as well as user reference material and descriptions of the operations.

Chapter 3: Practical Procedures for SCADA Software Development

This chapter offers a guide on how to design and develop the software for a typical SCADA project. From identifying the physical inputs and outputs of each controller, to the creation of the software documentation, to the final testing and checkout of the SCADA system, this chapter presents methodologies which have been used many times on many different projects. Each of the procedures is then covered in detail in following chapters.

Chapter 4: Documentation for SCADA Systems

This chapter presents a detailed explanation of the documentation that every SCADA project should have. Beyond the controller program listings, this chapter explains the purpose and content of the documentation that is produced as a result of good development procedures, as outlined in the previous chapter. Each of the documentation sections is then addressed in detail in subsequent chapters.

Chapter 5: Tagnames and Signal Naming Conventions

This chapter offers a system of naming database points which allows for a structured approach to tagnames. A tagging system is needed to establish a unified naming convention for both the field I/O signals and the internal program points used in the software. The method presented can be modified and tailored to the reader's specific application.

Chapter 6: Developing the Application Program Databases

This chapter presents a methodology of using spreadsheets to organize and document all of the points in the databases for both the field controller and the user workstation software. By applying the unified tagging system presented in the previous chapter, a consistent database can be created for all parts of the application software. Once the project has been completed, these same spreadsheets serve as database documentation.

Chapter 7: Process Control Logic Descriptions

This chapter provides a detailed explanation of every operation performed by every application program in every field controller. As this document, together with some of the initial database spreadsheets, form the design documents, a complete design document set results that can be used by the application programmer to develop the actual programs for the field controllers. These documents also serve as 'Shop Drawings' in contracts, as they clearly explain the design intent of the software before the detailed programming begins.

Chapter 8: User Operations Reference Manual

The manual in this chapter is needed for the user of the system; that is, for the plant operators and supervisors. Sections in this document describe the process operations available to the user, including the control functions and setpoint parameters that can be entered. Historical trends and reports are explained including how to extract data from a trend into a spreadsheet file for transfer to other parties. Any operation that a user can perform and any alarm or abnormal condition that may arise are addressed in this manual.

Chapter 9: Guidelines for Controller Application Programming

This chapter presents structured approaches to organizing the program logic for the field I/O controllers. While the logic can be kept entirely in one program file, a modular approach using subroutines results in a better organized and easier to understand program. Controller software today allows for not only multiple routines within a program, but also multiple tasks, each with multiple programs and routines. This chapter presents an approach that can be applied to almost any SCADA application, making the most efficient use of the project structuring features of the software.

Chapter 10: Guidelines for Workstation Application Programming

This chapter is similar to the chapter on the controller, but addresses the design of the Human–Machine Interface and the organization of the process and historical databases, along with many other aspects that relate to a well-designed user interface. The overall structure of displays, colour conventions, alarm handling procedures and trend displays are addressed.

Chapter 11: System Integration, Commissioning and Checkout

This chapter offers procedures or methods to implement application software for a SCADA system in a stepwise manner. Once the software has been developed, it is ready to test. But often the actual field controller has not been fully wired and/or all of the process equipment has not been installed and checked. The application software is always the last part of a project to be implemented, since it requires all of the components of the system to be installed and tested. The approach in this chapter presents a method that allows for the incremental testing and checkout of the application software.

Chapter 12: A Sample Project – Applying the Principles

This chapter presents the development and design of a complete SCADA application, including the field I/O controller program and the SCADA workstation software. All of the concepts presented in the book are applied to show how all of the material presented can be applied to a typical programming project.

Appendix A: Glossary of Technical Terms

This glossary offers a summary of acronyms and terms used throughout this book. Most of these terms are common in the SCADA systems industry.

Appendix B: TSNC Dictionaries

The Tagname and Signal Naming Convention described in Chapter 5 is illustrated here with a sample set of tagname fragments from which tagnames can be constructed. The tagname system proposed uses a series of five fragments of phrases to describe each part of the tagname; example fragments are listed here for a water treatment plant, to serve as an illustration of the tagname dictionaries referred to in Chapter 5.

Appendix C: Sample Process Control Logic Description

A sample PCLD is provided here to illustrate the application of the concepts presented earlier in Chapter 7, showing a real-world example for a sample PPC. All of the sections of a PCLD are included here for this one PPC system.

Appendix D: Program Listings for PPC Program

The program listings for the sample project are included in this appendix; the project is described in Chapter 12 for a pumping station. All of the program logic for all routines are shown, to provide a real-world example of a complete PPC control program.

Now that the overall content has been described, please read on to learn how to design, document and program the various parts of today's SCADA systems.

Stuart G. McCrady

1 Introduction

Objectives

- Explain the origin of SCADA
- Summarize the processing of input/output signals
- Identify the scope and function of the software in SCADA systems
- Explain the importance of documentation for software systems
- Highlight two major reasons for the development of this book

Today's Supervisory Control and Data Acquisition (SCADA) systems incorporate Programmable Logic Controllers (PLCs), Human Machine Interface (HMI) workstations and network communication systems into a complete integrated system. Each of the major components requires one or more form of programming from program logic to configuration to process graphic displays to communication configuration. Since there are many aspects to the SCADA application software, it is important to use a structured organized approach to the design and development of this software; hence, the establishment of standards is very important.

1.1 SCADA: Convergence of Evolving Technologies

Before identifying and discussing the elements of SCADA software, let us first take a brief tour of what SCADA is, where the term came from, and how it came to be. The term means Supervisory Control and Data Acquisition. The term was first used in the late 1980s but did not become a common term until the 1990s, as the technologies to be discussed evolved. There have been Distributed Control Systems (DCSs) since the 1960s and 1970s, but they were tightly integrated and proprietary control systems confined to a single plant or facility.

1.1.1 Early Automation Systems

In the 1960s and 1970s, the minicomputer served as the predecessor to the current day PLC; the minicomputer was programmed in assembly or machine language and interfaced directly with plant devices. The minicomputer was a general purpose digital computer, capable of being programmed in a few different languages, but it was primarily used for automation systems in plants. I have programmed many such systems, mostly in machine language, but some even used FORTRAN and ALGOL!

The PLC first came into being around 1971, designed and built by Gould Modicon, and was intended to replace the traditional relay ladder logic electrical

Designing SCADA Application Software. DOI: http://dx.doi.org/10.1016/B978-0-12-417000-1.00001-4

circuitry. The programming language was developed to mimic the relay logic of the electrical circuits, with 'power' and 'return' rails on the left and right, respectively. By around 1977, Allen–Bradley introduced their first major PLC, by which time the idea of a programmable controller had become quite widely accepted. The minicomputer was being replaced with these new programmable devices, primarily because the language was already familiar to the electricians, and hence learning to 'program the circuits' was pretty straightforward. These controllers were also digital computers but were designed to interface to the basic field signals we still use today: discrete inputs and outputs, and analog inputs and outputs.

The PLC hardware and software continued to develop and evolve until the programming became much more than the original electrical circuitry that they were meant to replace. With the introduction of Microsoft Windows and the Graphical User Interface (GUI), more intuitive and graphical programming became possible. Today, the PLC programmer expects an easy to use yet feature rich programming environment.

1.1.2 The Human Interface

Back to the 1970s, the user or human interface to the early minicomputer systems consisted of a monitor and a keyboard, with alphanumeric text only. Getting the computer to display a line of text was an accomplishment, and when colour came along, 'special effects' were possible; for example, having different coloured text and some basic shapes like a square could now be produced. Once again, the introduction of Windows and the graphical interface made many things possible, not just coloured text. By the late 1990s, extensive process control graphic displays were possible with animation and graphical representation of field conditions.

The first personal computers were developed for commercial use by IBM, and companies like Intellution developed complete user interface software packages which executed on the personal computers, and could be interfaced to the PLCs. This SCADA software provided a graphical interface into the system, as well as, both current and historical database files which could be accessed for reporting information. Other companies developed personal computers, of course, and today there are many different brands available, all of which can serve as a user interface workstation to the process world of the plant or system.

1.1.3 Communications and Integration

However, the communication technology had to develop in order to integrate or combine the software on the personal computer with the programmable controllers in the plant. Initially, the communication involved proprietary protocols between the computer software and the PLC, but now Ethernet with TCP/IP has become one of the industry standards. The personal computer software, which became known as the HMI software, could then be integrated seamlessly with any of the brands of PLCs on the market. High-speed networks and communication systems are standard today, but in the 1970s and 1980s, a dial up modem with 1200 bits per second was considered great!

So, by the 1990s, all of this technology had developed to the point where almost any PLC brand and any HMI software brand, could be integrated into a system of any number of workstations and PLCs. By this time, the term SCADA had become a common term, even though there were automation systems using minicomputers and custom-designed hardware interfaces in the 1970s, which were the early versions of SCADA systems.

A SCADA system, therefore, can be considered any combination of PLCs and HMI workstations, supported by a communication network for both in-plant facilities and remote sites that functions as a fully integrated system. The communication facility now extends to the Internet for Wide Area Networks (WANs) and regional systems.

1.2 Basics of SCADA Signal Processing

Before getting into any details on the software involved in SCADA systems, it might be helpful to review some basic concepts on how field signals are processed and interfaced with the field controllers, such that the application software can work with and interact with the physical electrical world for which these controllers were designed. The field controller, commonly referred to as a PLC, is a specialized computer which works with binary information. The purpose of the input and output modules of the PLC is to convert signals from the physical world to and from the internal software world.

There are only four basic data types which are connected to the PLC: discrete input, discrete output, analog input and analog output. All input signals processed by the PLC and all output signals driven by the PLC are in one of these four forms. Hardware modules in the PLC chassis or rack use electronics and firmware to process the input signals and generate the output signals.

The discrete input is a two-state signal, represented by electricity flowing (True) or not flowing (False). The binary system relates this to the '1' and '0' values of software. A discrete or digital input might be a switch, photocell, pushbutton, contact, or proximity sensor; the signal is either on or off. The discrete input module would represent this presence or absence of the electrical signal as '1' or '0'.

The discrete output is a two-state signal, represented by electrical flow also, but is sent to devices such as lamps, motor contactors and solenoid valves. The '1' or '0' state in the internal software is converted to the presence or absence of an electrical signal to the device; the electrical circuit is either on (1) or off (0). Hence, devices can be turned on or off, depending upon the state of the binary point in the PLC program.

Analog inputs are represented by a range of electrical signals, such as 1–5 Volts Direct Current (VDC) or 4–20 milliamps (mA). This electrical signal is generated by a transducer, which converts a field value to a proportional electrical signal. The analog input module samples the input signal, and converts it to a 16-bit binary number in the range of 0–32,767. The low end of the signal, 0 VDC or 4 mA, would be stored as a value of zero (0); the high end of the signal, 5 VDC or 20 mA, would be stored as a value of 32,767. The current value of the voltage or current is converted to a binary number which is proportional to the electrical signal between the two limits mentioned. Hence, the PLC works with binary numbers in the range of 0–32,767, representing the field signal of 1–5 VDC or 4–20 mA.

Analog outputs begin as internal 16-bit values, and are then converted to an electrical signal in the range of the 1–5 VDC or 4–20 mA; these signals in turn drive speed controllers, valve positioners, and any other variable control device. The output signal is in the same proportion as the internal 16-bit value. So the analog output works the same as the analog input signal but in reverse.

The field controller therefore works entirely in software using the binary number system. Discrete and analog data are represented by either a single binary digit, or a group of 16 bits called a word or integer. In this binary form, programs can then perform calculations and comparisons to perform control operations.

Working with analog data values, for example, turning an oven on or off can be determined based upon the temperature of the oven; following is a structured programming version of the logic involved:

```
IF Furnace.Temperature > High.Temperature.Limit
  THEN
      Turn off the heat to the oven.
  ELSE
      IF the Furnace.Temperature < Low.Temperature.Limit
  THEN
      Turn on the heat to the oven.
  ENDIF
```

Working with discrete data values, a motor start/stop control output can be turned on or off based upon the required states of specific input signals, as shown in the following programming sample:

```
IF Motor.Remote = True .AND. Motor.Alarm = False .AND.
Motor.Start = True
  THEN
      Turn on motor run control output.
  ELSE
      Turn off motor run control output.
  ENDIF
```

Throughout this book, there are references to discrete and analog data; it is important to understand where this data comes from, and to remember that input and output modules are used to convert or translate between the physical or electrical world and the internal software world of binary numbers. A discrete input or output is the presence (on) or absence (off) of an electrical signal, while an analog input or output is a varying signal between two limits, such as 1–5 VDC or 4–20 mA.

1.3 Defining the Scope of SCADA Software

Traditionally, the PLC was programmed by electricians, as the ladder logic programming was designed to resemble electrical diagrams. At first, this approach worked well, as the electrician already understood how a program should work. But as the PLC software evolved into more complex features and operations, the programming extended well beyond what could be done in an electrical wiring diagram. With this added functionality, there was a need to better organize and design the software for the PLC.

At the same time, HMI, developed from simple meters, chart recorders, and push-button and selector switches, into very sophisticated graphic interfaces. First there was the Cathode Ray Tube (CRT), which was combined with a keyboard. Then the CRT led to Liquid Crystal Displays (LCDs), which required much less space with better resolution. As this user interface evolved, there was again a need to better organize and design the images and information being displayed to the user.

With both user interfaces (i.e. HMI) and sophisticated field controllers (i.e. PLC), the combination of these devices evolved into what is now referred to as a SCADA system. Interconnecting the HMI and PLC equipment required more advanced methods of communication, such as networks and special driver software. And then came the issue of interconnectivity with other systems: the result was Local Area Networks (LANs) and WANs.

SCADA software involves much more than a set of engineering drawings; as will be explained in subsequent chapters, the application software for a SCADA project involves spreadsheets, design documents, user reference manual material and detailed program information. Thus, the software for these systems is very extensive and requires a combination of design documents.

Today, designing, developing and implementing SCADA systems requires a number of areas of programming, each with its own special requirements and languages. As will be explained in the next chapter, there is programming required at all levels of a system. Understanding what these levels are, what each one requires and how to approach the design requires a much more methodical approach to programming than ever before.

1.4 Use of Generalized Terminology

Perhaps the most common names used for equipment in the SCADA system are the PLC and the HMI. While these terms are well known, for the purpose of this book some more general names will be used. Since some manufacturers use some different terms, I have chosen to use some generic names. Refer to the Glossary in Appendix A for a complete list of acronyms and technical terms used in the SCADA systems programming field.

Since the term HMI refers to the process graphic interface, a more general term will be used for the workstations. The PLCs vary in the level of functionality, so a more generic name will be used herein. And the tagging system used for naming both field input and output signals, and for naming internal program points, will be Tagname and Signal Naming Convention (TSNC).

A few other commonly used terms in this book are included here, as well as in the Glossary. These terms and/or acronyms are presented here for introductory purposes.

PPC **Programmable Process Controller**
Specialized computer programmed in one or more languages to effect control and to monitor field conditions; other acronyms include: RTU, RPU, DPU, PLC, PAC

SOW **SCADA Operations User Workstation**
Computer executing SCADA software which includes the HMI process graphic displays, process and historical databases, and background process operations; other acronyms include: HMI, View Node, SCADA Node, SCADA Host Workstation

HMI **Human Machine Interface**
The operator interface, in the form of process graphic displays and historical trends, constitutes the HMI; however, the HMI is one part of the overall SOW application

OIT **Operator Interface Terminal**
Small display panel with minimal keys for displaying local information for a PPC and entering/changing values for the local process

TSNC **Tagname and Signal Naming Convention**
Standardized naming convention for assigning names to the various field I/O signals as well as the internal programming points

PCLD **Process Control Logic Description**
Standardized documentation for the detailed explanation for the process operations of the PPC application software

1.5 The Need for Programming Standards

The life of a SCADA system project outlasts the original programmers, so that future additions and changes typically require someone new to work on the software. It is not unusual to find a SCADA system that has been in operation for 10–20 years, as the equipment generally has very long lifetimes. The technology of the PPCs today allow these controllers to last 30 or more years; the principal reason they are replaced is that a better, faster, cheaper product becomes available. Likewise, once the application software has been developed and commissioned, changes and additions are made over time. So the main reason for replacing any of the SCADA equipment is to update with faster and better technology.

1.5.1 Need for Design Standards

As a result of the long life of SCADA systems, there could be many programmers involved in the development of the application software. At the start, the initial programmer or programming staff design, program and implement the application software for the SCADA project. Then over time, additions and changes are made but often by another programmer. Given enough time, there could be three, four or more people involved in the programming of the system. Since programming is still very much an art, and each person has his own particular style of programming, there could be significant differences in the way that each new person wants to program the system. It is not unusual for a new person coming into the project to choose to redesign and reprogram the application in order to implement his own style, as his style differs considerably from what has already been done.

In order to avoid having each new person reprogram the system to his style, there needs to be some programming standards and conventions implemented at the start. Issues such as tagname conventions, PPC program structure and format of HMI displays must all be considered up front before anyone begins programming. By establishing standards on what the finished product should look like, and how the software is to be organized and structured, each new programmer can still use his programming style but work within the framework of the original system design. If changes are to be made to one of the PPC control programs, the overall structure and organization has already been established, so the new programmer can quickly identify how his new efforts will dovetail or fit into the existing software. In fact, by following this approach, this actually saves the new programmer considerable time, since the general design structure has already been established, and the programmer simply needs to find that part of the application for which the changes are required.

1.5.2 Modify or Reprogram

As a result of the longevity of SCADA systems, changes and additions are often made long after the original system was installed; here again the adherence to standards in design will make these changes simpler, as the programmer can refer to the existing design documentation and become familiar with the programming more quickly.

I personally have been hired to modify an existing PPC application program, only to discover that the lack of proper design and documentation in the original work necessitated me reprogramming the project from scratch. In the end, I not only implemented the required changes, but I also produced the important design documents that I felt should have been there from the start. Of course, I worked closely with the end customer to ensure that I fully understood the original design intent before proceeding to develop the new programming work.

An expression I learned many years ago states: 'The good thing about standards is that there are so many to choose from!' So it is important to choose only one, and then stay with it, regardless of who implements the application programming work.

1.6 The Importance of Software Documentation

As indicated in a previous section, the complexity of application software for SCADA systems has increased substantially over the past 40+ years. Automation programming is no longer limited to a PPC with an OIT. In the next chapter, the various software application areas of a system will be examined, and their significance within the SCADA project will be identified.

With so much software, there is a real need for proper documentation for all aspects of the application. This includes documentation on the tagnames and databases used at the different levels of the system; documentation on how the automatic programs function, including their interfaces to both the operator workstations and to other process controllers; documentation for the operations personnel on exactly what and how the user is able to do and what limitations are in effect due to security levels; and of course the controller program listings themselves.

1.6.1 Lifetime of Software Systems

The hardware used in SCADA systems continues to be more reliable and longer lasting. Once a SCADA system has been commissioned and is in regular use, the equipment long outlasts the original person or people that designed and programmed the application software. While it might be possible for one person to 'know' the entire system at the time of developing and commissioning the project, years later it would be difficult for that same person to still know the system well enough to implement changes and additions. Without good software documentation, then, consider the challenge facing someone new coming in to make changes to the existing software. Since the system integration people who typically implement these systems often change jobs and/or move to other locations, the end user or customer is left with a system for which they have very little information.

If another person is brought into the picture to modify and add to the system, this new person needs some form of design documentation to reference. It is perhaps more for this reason than any other that having a properly documented SCADA project is so very important. The new person involved in the SCADA system would require considerable time to 'get up to speed' on how the current system is designed, before being able to make any changes or additions without adversely affecting the operation of the current system.

1.6.2 Upgrades and Revisions

Finally, consider the case of upgrading all of the equipment in the system to newer hardware and software; during the life of a typical SCADA system, a point is reached at which new additions and/or modifications are simply not possible, as the current hardware and software cannot accommodate the change.

New application software for both the SOW and the PPC are often required; with new versions of development software, there is often a need to upgrade other software, such as I/O servers and drivers, and even the operating system (e.g. Windows XP to Windows 7). Existing application software developed for the workstations and the controllers can usually, but not always, be imported or converted to the new development software. Rockwell Automation, for example, includes a conversion utility for converting PLC-5 and SLC-500 applications to the newer ControlLogix platforms. The database structure in the ControlLogix system is significantly different so the converted data points are assigned an older style of 'tagnames'; the programmer must then go through the entire application to create new meaningful tagnames based upon the previous point names. For example, a Boolean flag for 'Pump Ready' might be 'B3:7/12'; the conversion utility might create a

ControlLogix tag of 'B3_7_12'; the programmer would then need to create a new tagname, such as 'Pump_Ready', then find and replace all references to the B3 type name to the newer text-based tagname.

If the original software system was properly designed, then revising the database point names could be much simpler. If the original system used a structured tagging system like the one described later in this book, then the new tagname could be constructed, and then the find and replace operation could be performed. An address such as 'B3:7/1' might have represented CL3_MTR2_00_SRN for Conveyor Line 3, Motor 2 running status; in the newer ControlLogix database, the new tagname would simply be the structured tagname.

There simply is no substitute for good documentation in SCADA systems.

1.7 Purpose and Overview of This Book

The purpose in writing this book is twofold: offer a practical approach to the design, development and implementation of application software for SCADA systems; and present a complete documentation system that addresses all aspects of a project.

1.7.1 Organized Design and Development

The practical approach to the development of application software presented in this book is the result of designing and implementing more than 50 such SCADA projects. Each project that I have completed allowed me to refine and improve my methods for developing the software in such a way that the design documentation ultimately became the final documentation for the project. The methodology presented herein also streamlines the development processes, saving time for the programmer. The end users or customers for whom I was contracted to develop the software all commented on the completeness of the documentation and the fact that anything that someone wanted to know was readily found in the material.

Future additions and modifications also proved to be very straightforward, whether I did the work or someone else did the work. In Chapter 4, there is a complete description of the software documentation that I consider constitutes a complete document set for any SCADA project.

1.7.2 Complete Software Documentation

The second reason for writing this book is to provide the reader with useful and practical information on how to prepare documentation for an existing system. This 'reverse engineering' would therefore result in documentation for the existing system, and make future work considerably easier. I have witnessed many existing SCADA systems for which there was no documentation; there were no logic descriptions to explain how the controllers performed their operations, no record of the databases used in both the controllers and the workstations, and no comprehensive user manual for the people responsible for operating the facility. However, if one could develop documentation consisting of spreadsheet points lists for all field controllers and a description or narrative explaining how the controller implements its logic, this would be a major step forward for anyone responsible for maintaining the system.

This book is about how to design and develop application software for a typical SCADA system. Throughout the book, I offer practical step-by-step procedures for creating the necessary software. The emphasis is on developing the software using documentation techniques which will result in a complete set of documents for every part of the application software. The style is that of simple guidelines which can be applied to any application of any sized SCADA project. The focus here is to provide the reader with a straightforward approach to organizing the SCADA software by process area, create the databases and application programs, and integrate and commission the entire SCADA system.

1.7.3 Adherence to Programming Standards

The IEC 61131-3 programming standard mentioned in the preface establishes a vendor independent programming environment for the PPCs of a SCADA system. Since the PPC application programming requires the most amount of time in a SCADA project, any savings in time has a direct impact on the cost of the project. The five languages included in this standard offer maximum flexibility to the programmer. In Section 9.3.5 on programming the PPC, the five languages of this standard are illustrated.

The PPC is the heart of any SCADA system, as these controllers monitor and control all operations in the system. The operator workstations provide 'windows' into the various processes, but the PPCs perform most of the work. This IEC programming standard allows the programmer to choose a language best suited for each part of the complete application; it is possible that in a single PPC program, all five languages might be used, with different sections being programmed in a different language. Referring to the importance of standards in programming and in documentation, this programming standard promotes good structured programming style, which results in application software that is easier to troubleshoot and commission, and is easier to modify and maintain in the future.

From my own experience in the industry, I have found that some parts of a PPC program are better done with Function Block programming, while other parts are best done with Ladder Logic. For sequential operations involving a number of steps, Sequential Function Chart may be the best choice. I have programmed systems which involved both the traditional ladder logic operations and data manipulation, or 'number crunching' as it might be called. By being able to select a different language for certain parts of the program, I have been able to develop the software more efficiently, benefiting from the features of each of the programming languages. As a result, my documentation was simpler, cleaner and easier to prepare.

Thus, the need for standards applies not only to the software design and documentation but also to the actual programming environment for the field controllers. Software developed following these standards is easier to understand and document, and easier to develop and implement.

2 The Elements of SCADA Software

Objectives

- Identify the major components of a SCADA system
- Illustrate common network topologies in use
- Describe and distinguish the programming requirements of the field controller and operations workstation
- Identify the communication methods used in SCADA systems

Programming software for Supervisory Control and Data Acquisition (SCADA) systems involves both SCADA Operations User Workstation (SOW) software and Programmable Process Controller (PPC) software. In addition, SCADA systems require programming of an interface between the user workstation computers and the programmable controller equipment. There are many considerations involved in the design and development of the application programming software required at the various levels of a typical SCADA system.

Consider a typical material handling system to illustrate the different processes involved in a SCADA system. A materials handling system, such as a beer bottling line, consists of stages or areas of operation, each of which can be considered one complete subsystem. The process operations are specific to that area, such as bottle filling, bottle capping, grouping for packaging and packaging into boxes. Within the subsystem, there are field signals to the PPC, control signals from the PPC, and communications to and from the SOWs. Users can interact with the system via the SOWs, issuing commands, adjusting setpoints and parameters, and responding to alarm conditions.

2.1 Understanding the Elements of SCADA Software

The typical SCADA system today consists of user operations workstations, SCADA server computers, communication networks, programmable field controllers, and field devices and signals. All of the components are seamlessly integrated into a complete operational system, which provides both automated operations as well as user-controlled actions.

Whether the SCADA architecture has already been defined, or it is undergoing detailed design, the software requirements of each process area must be considered. There are field devices which provide signal data to be processed by a PPC; there are major equipments, such as pumps, motors, valves, which must be monitored and controlled. The detailed operation of each PPC must be described so that the design

Designing SCADA Application Software. DOI: http://dx.doi.org/10.1016/B978-0-12-417000-1.00002-6

can be translated into an application program. And the operator workstation must have graphic displays, trends and alarm/event summary reports to operate the complete system.

The goal of subdividing the entire system into smaller more manageable subsystems is to identify functional areas, each of which can be designed to perform area-specific operations. All of the signals or data required for the area, together with the program logic required to effect the required operations, must be clearly defined and documented.

2.2 Typical SCADA System Architecture

The architecture shown in Figure 2.1 illustrates a typical SCADA system in terms of the components or elements which are interconnected via a communication network. Following the illustration are more detailed explanations of each of the major components.

In summary, there are five major areas or components to the SCADA system, as listed below and described in more detail:

- Field Devices and Signals
- Programmable Process Controller
- SCADA Operations User Workstation
- SCADA Server Computer
- Communication Network

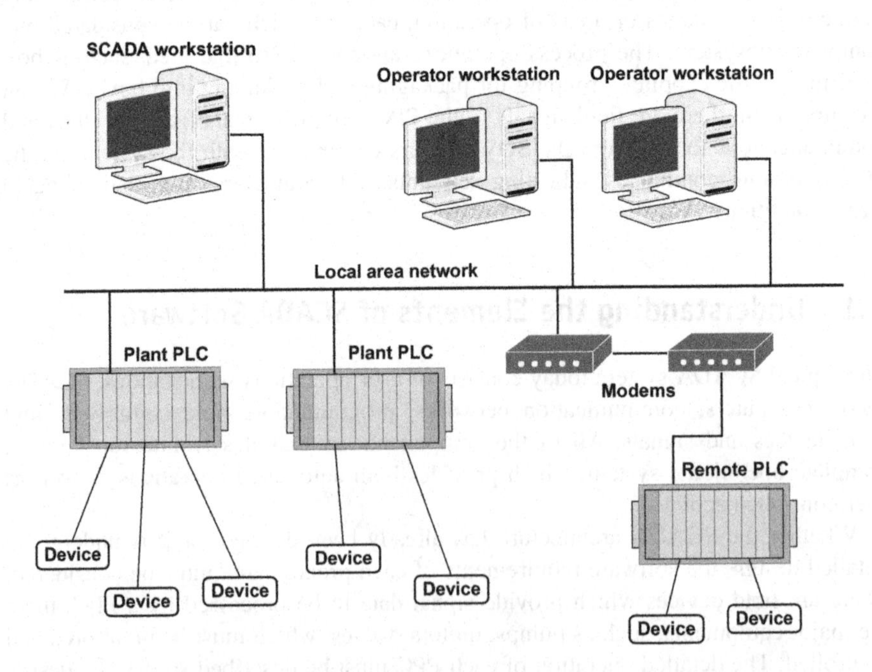

Figure 2.1 Typical SCADA system architecture.

The Field Devices are represented as signals into and out of the PPC; this includes controlling devices, such as pumps, valves, solenoids, etc. These field devices are the link between the SCADA technology and the process operations of any automated facility.

The PPC is the heart of any SCADA system, as the PPC performs the monitor and control functions over the field devices. Programming software for these components can become quite complex with multiple tasks, programs and routines.

The SCADA Operations User Workstation provides the user interface in the form of process graphic displays, trends and associated reports. Most SCADA systems have multiple workstations through which the user can interact with the system to monitor operations and effect equipment control.

The SCADA Server Computer is used to maintain the process and historical databases for the user workstations, as well as provide the communication interface between the server computer and the PPCs.

The Communication Network is the hardware and software that interconnects all of the components of the system. Typical networks today include Ethernet with Transmission Control Protocol/Internet Protocol (TCP/IP) and proprietary communication topology.

2.2.1 Field Devices and Signals

The icons in the illustration labelled 'Device' represent individual signals; these may be discrete or analog, input to the PPC or output from the PPC. When designing the software for a SCADA system, the various field signals and/or devices must be considered in terms of what information is to be monitored and what equipment is to be controlled.

Field devices may be signal transmitters, such as level or pressure transmitters; they could be discrete signals, such as a valve's open or closed status, or the motor's running status. Some devices may actually provide multiple signals, such as a water quality unit which provides both chlorine residual and water pH. When designing SCADA software, all of the field signals must be identified for each process area. As shown above, the field devices are wired into the various PPCs, which in turn process the signals.

All of the field devices, which are represented by discrete and analog signals, are connected to the various input and output modules of the PPC. The PPC processes the input signals and effects control through the output signals, based upon the programming in the PPC.

Discrete signals can be both input and output types; such signals are two state, meaning the signal can only be in one of two states at any time. Discrete input signals may include: open and close limit switches, photocell sensors, pushbuttons and selector switches. Discrete output signals may include pilot lights, motor control relays, solenoid valves and valve controls.

Analog signals can also be both input and output types; such signals can have a range of values between two preset limits, most often zero and some maximum value. Analog input signals may include levels, flowrates, motor speeds, voltage and current from power monitors, water quality signals, and temperatures and pressures. Analog output signals may include motor speed controls for Variable Frequency Drives (VFDs) on variable speed motors, valve positioning signals for modulating valves and analog display devices.

Since field signals, such as levels, flows and pressures may be used by the program to control the operation of the equipment, the Programmable Logic Controller (PLC) is the heart of the monitor and control of every process area in a SCADA system.

SCADA software must consider all of the equipment and signals to be processed. For example, consider a Raw Water Pumping Station for a Water Treatment Plant (WTP); there will be two or more pumps, valves and field signals. For each device or equipment, one must identify what signals are required both from the equipment (inputs) and to the equipment (outputs). Hence, a unit of equipment may be represented by a collection of signals connected to the process area PPC. A pump, for example, could include multiple input and output signals.

In addition, there are typically field signals which are required for the proper operation of the process area. Again for the WTP application, a raw water pump station might include intake well level, water turbidity, water pH, water temperature, and discharge flowrate and pressure. These separate field signals will be used by the application program executing in the PPC.

When complete, the field device and signal list for any process area should include all of the signals related to both the physical equipment and to the raw signals. As will be explained in more detail in chapter 6, 'Developing the Application Program Databases', a detailed signal list will summarize all information for each process area.

2.2.2 Programmable Process Controller

A given application for a SCADA system is divided into process areas. Each such area has clearly defined operations to be performed. For example, a pumping station may use two or three pumps, operating in a lead/lag/standby mode; the automation program in the controller is configured to operate the pumps based upon operator-entered setpoints and duty assignments.

Each process area identified within a SCADA system will require its own PPC. The PPC will be programmed so as to monitor all of the signals within the process area, and to effect control over the process equipment, based upon the design of the program.

The software designer must allocate the various field I/O signals identified in the previous topic to the required modules in the PPC. With the list of signals, organized into discrete input, discrete output, analog input and analog output, a count of the number of each type of signal can be determined. From there, the necessary input/output modules can be determined and configured for the PPC. Depending upon the number of signals, a PPC may consist of multiple racks or chassis of modules. From the perspective of the application program, however, the software recognizes all modules as though they all reside in a single long rack.

Each of the programmable controllers requires programming in one or more forms. For example, the field controller, commonly referred to as the PLC, is typically programmed in a language called Ladder Logic, which resembles the electrical control circuitry used before the PLC came into being. Today, most PLCs allow for a variety of programming languages to meet the application requirements; for example, languages include: Function Block Diagram, Structure Text (High-Level Programming), Sequential Function Chart, and Instruction or Statement List (Low-Level Assembly Programming). The programmer may choose to use one or more of these languages in a specific PPC application program.

2.2.3 SCADA Operations User Workstation

The user operations workstations, usually referred to as the Human Machine Interface (HMI), requires the programming of process graphic displays with animated links to many points in a process database. Configuration programming is also required to establish the process database, the historical database and the communications interface to the field controllers or the PLCs. Additional background programs, called scripts, are often used to perform 'behind the scenes' operations for the application.

The SCADA workstations present the process graphic displays with operator interaction, such as controlling equipment and requesting information. The software at this level involves the creation of the process control displays, historical trend and historical report displays, alarm and event summary displays, and the process database. Hence, there are both displays which present the information and 'behind the scenes' programming to access the information for the requested display.

In addition, there are often background scripts or programs that are used to perform operations associated with the displays and/or invoke commands to the PPCs and other equipment. Such scripts and background programming will be addressed in later sections.

2.2.4 Communication Network

The in-plant equipment, PPCs and SOWs, are typically interconnected via a Local Area Network (LAN), using Ethernet or other high-speed communication system. Some SCADA systems may extend outside the physical building into remote sites; these sites require some form of communication back to the host facility also. The illustration of the SCADA hierarchy shown previously includes modems to a remote site, allowing remotely located controllers to operate over the same high-speed communication network.

The organization and/or topology of a SCADA system is beyond the scope of this book, but I felt it was worth including a brief explanation of the types of system architecture, or network topology, used for SCADA systems. There are three basic topologies as described below.

The *Bus Topology* as shown in Figure 2.2, consists of a hardware/software interconnection among all of the nodes in the system. This architecture resembles a major

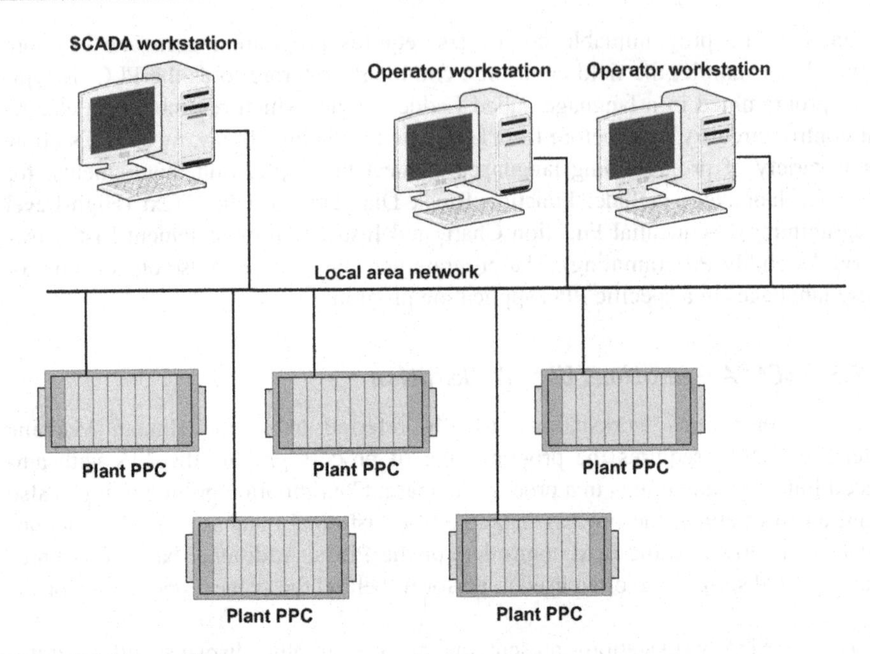

Figure 2.2 Bus topology.

roadway to and from which all other roads connect. To travel from any one location to any other location requires getting on to this major roadway (network) and then travelling along until an exit to the desired route is found. All traffic or communications in the system is accomplished via this single bus-type network. For increased traffic, the network can become overloaded, and the result is a slowing down of the transfer of data from one node to another. While Ethernet over a bus network is generally fast enough, there may be some applications in which this bus topology creates roadblocks to efficient data.

The *Star Topology* as shown in Figure 2.3, consists of multiple network paths out from a single master or host node. This master node would typically consist of one or two master SOWs functioning as masters of the system. All data collection from the various PPC nodes is done through individual connections in a star configuration. Update times to the host node are very fast but does require multiple paths out from the host master node. The transfer of data between nodes on the Star network does require that the information be passed first from the source node, then through the host node, and then out to the destination node.

A *Token Ring Topology* as shown in Figure 2.4, works like a ring in which all nodes are interconnected by two network connections. All nodes in the topology are of equal value, and data is passed via this ring from one node to the next. Duplicate rings in opposite directions provide redundancy and security. Information from any node passes along the ring, being passed from node to node, until the data reaches the desired destination node. This topology is predictive in that the speed is constant and the time to transfer data is always at a fixed rate. As the number of nodes in the network increases,

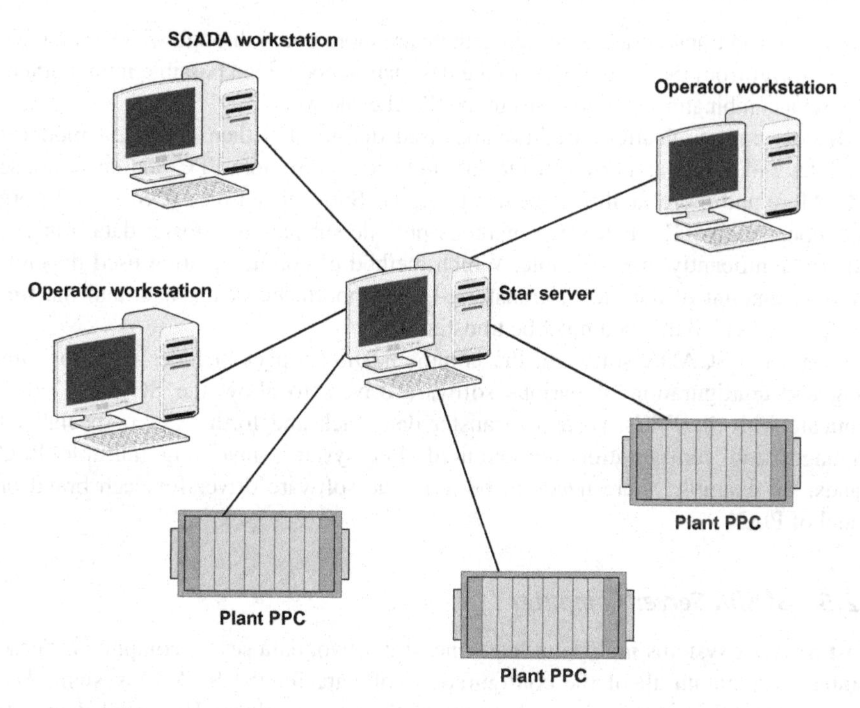

Figure 2.3 Star topology.

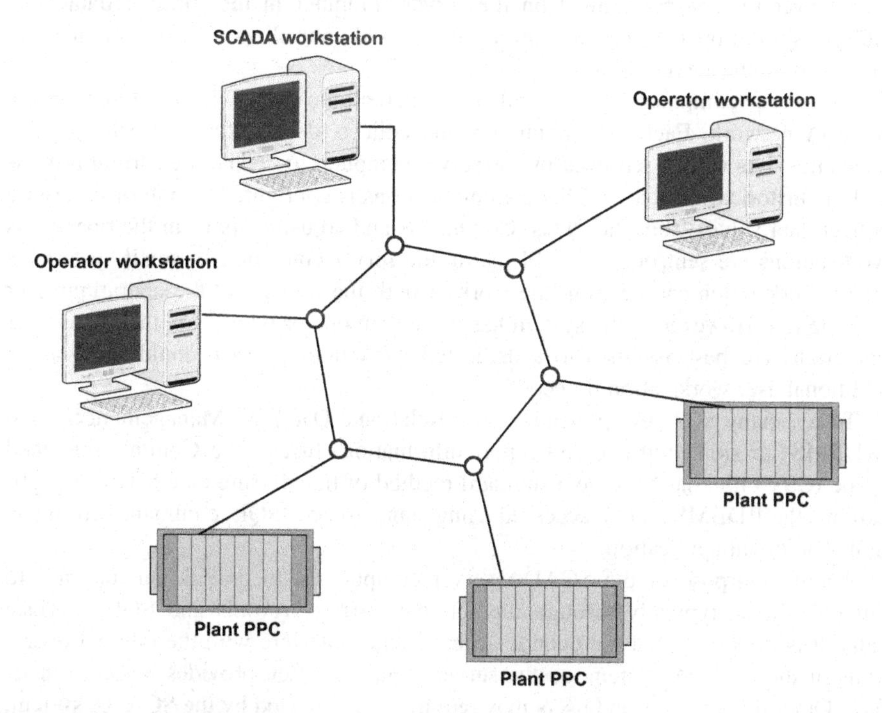

Figure 2.4 Token ring topology.

the overall data transfer rate drops since there are more nodes through which data must pass to travel from the source node to the destination node. One possible improvement is to use a combination of topologies in the SCADA network.

Remote communications traditionally used dial-up and then dedicated modems (modulator–demodulator) to transfer data between the remote PPC and an in-house PPC. More and more facilities are using Digital Subscriber Lines (DSLs) and fibre optic connections. The latter two methods provide substantially faster data transfer, and are significantly more reliable. Which method of communication used depends upon the amount of data to be transferred, the importance of the data and the frequency with which the data must be transferred.

Within the SCADA software, the communication aspect includes the programming and configuration of various software drivers to allow the SOWs to communicate with the PPCs, so as to transfer data back and forth. Again, specialized languages and configuration are required. For systems involving multiple PLC brands, for example, there needs to be a unique software driver for each brand or model of PLC.

2.2.5 SCADA Server Computer

Most SCADA systems include at least one, if not two, data server computers. These computers maintain all of the configuration software for the SCADA system. The server computer is at the physical centre of the Star topology. Historical data collected over time is maintained on the server computer in the form of databases. Current system operating data from all of the field controllers is also maintained in databases on the server computer.

The server computer performs all of the communications with the PPCs on the SCADA network. Each PPC maintains and collects data pertaining to its process areas; this data is then retrieved by the server computer to update the current process and the historical databases. This communications is configured to poll or otherwise collect data values from the PPCs. Commands and adjustments from the operations workstations are sent out to the PPCs via the server computer. In small systems, a single workstation can perform the work of both the server and the operations user workstation. However, if the system has more than one or two PPCs, then the server operations are best assigned to a dedicated workstation, which could serve as an additional user workstation if needed.

Today, many SCADA applications use Relational DataBase Management System (RDBMS) to store, retrieve and report information. Just as the Comma Separated Value (CSV) file, has become a standard method of transferring data between applications, the RDBMS can be accessed using standardized 'SQL' commands from any SQL-compliant application.

Another purpose of the SCADA server computer is to provide an interface to other facilities, typically through the Internet, using Firewalls and SQL interface calls. It is important that the outside access cannot interfere with the internal operations of the SCADA system, so the server computer often provides a secure interface. Other departments and users may require data collected by the SCADA system, and so a means of accessing this data can be provided through the SCADA server computer, with the appropriate security measures in place.

2.3 Sample Application: WTP SCADA System

The best way to illustrate concepts and ideas about SCADA software is to use real-world applications; I have worked on more than 50 such SCADA projects, including more than two dozen water and wastewater applications. For the purpose of illustration of concepts in this book, a WTP will be used, showing how the information presented would be applied to actual programming applications in this field.

WTPs utilize the same automated technologies that material handling and automotive assembly lines use; PLCs for the field level monitor and control of operations, HMI workstations for the user interface and communication networks for interfacing all elements. In this illustrative application, the basic operation and SCADA elements of a WTP will be provided, focusing on the SCADA software involved in each area of the plant.

The general process flow and major equipment used in a WTP is illustrated in Figure 2.5; a brief description of operations follows.

The WTP generally consists of five process areas: Low Lift or Raw Water Station, Pretreatment, Filtration, High Lift or Treated Water Station, and Chemical injection systems. The Low Lift pumps the water from a source into the Pretreatment facility; there the raw water is treated chemically to remove bacteria and other foreign content. The Filtration system allows the water to pass through several layers of material to remove all undesirable materials and bacteria, with the effluent being potable water. The High Lift pumps then discharge the clean water out into the distribution system.

2.3.1 Low Lift or Raw Water Station

This section of the plant consists of an intake well into which raw water flows, and two or more raw water pumps. The pumps may be controlled both in Remote Manual mode from the SCADA Workstation and in Remote Automatic mode in which the PPC controls the pump operations. A flow control valve may be used to control the flowrate of water into the pretreatment facility. A constant flowrate into the rest of the plant is most desirable.

2.3.2 Pretreatment

The pretreatment section is the area in which chemicals are added to treat the water and destroy bacteria in the water. With proper treatment, the water quality leaving this section is already relatively clean. The various chemical pumps are controlled via the PPC control program to be paced to the raw water flowrate in and automatic mode of control. Often, the water quality leaving the pretreatment stage is substantially cleaner but not ready for drinking or consumption.

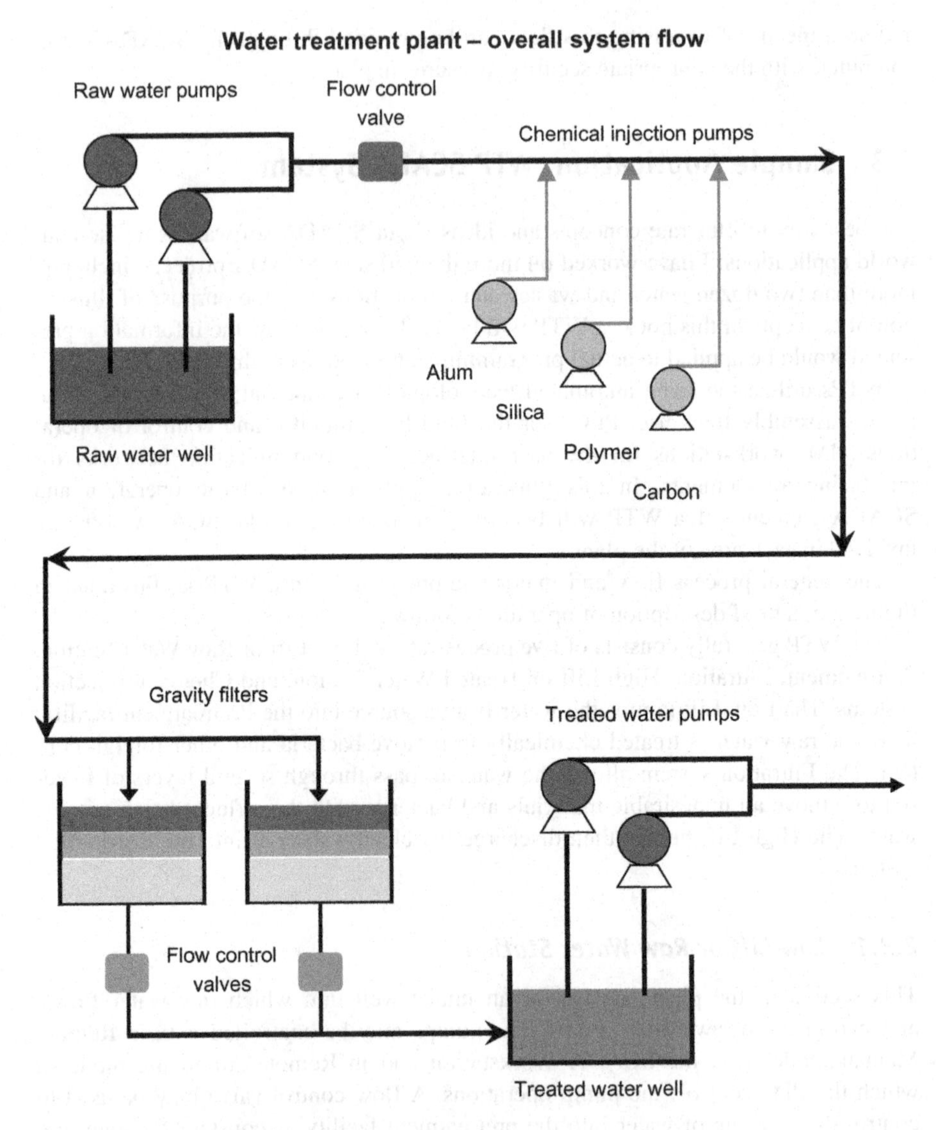

Figure 2.5 Process flow for a water treatment plant.

2.3.3 Filtration

The filters of the plant perform the major filtering or cleaning of the water. Typical filters consist of multiple layers of material, starting with granulated carbon on top down to concrete blocks at the bottom. Each layer serves to remove further dirt and bacteria from the water. The primary control here is the flowrate control through each filter; the flowrates are entered by the plant operators and used in the PPC control program. Flowrate control through the filters is typically achieved using Proportional Integral Derivative (PID) control logic in the PPC application programs.

2.3.4 High Lift or Treated Water Station

This final section pumps the treated water from a clearwell reservoir out to the water distribution system. Pumps are controlled based upon an external level signal from an elevated water tank and/or from the plant distribution pressure. The plant operators enter operating setpoints for the automatic pump operations. Chemicals may be injected as the water leaves the facility to ensure the water remains clean and drinkable, and to ensure that the quality remains within limits established by the Ministry of the Environment.

2.3.5 Chemical Systems

In the treatment of water, various chemicals are injected throughout the process to remove bacteria and other harmful elements of the water. Most chemicals are injected near the start of the process, such as the pretreatment stage, to purify the water before filtration. As the treated water is being pumped out into the system, additional chemical may be injected; this might include chlorine to ensure the treated water remains drinkable as it travels through the many water mains of a city or town.

The chemicals are typically injected in a flow-based approach, that is, as the raw water flowrate changes, the amount of chemicals change. This ensures that the water maintains a constant level of the required chemicals through the entire process. PID control systems, implemented in the PPCs, are normally used to control the injection of chemicals based upon an operator-entered dosage setpoint. Thus, the chemical systems are an important part of the overall water treatment process.

2.4 Getting the Most from Field Data

SCADA systems are designed to monitor many different field signals and effect control over multiple processes in the system. The data being monitored consists of both discrete (i.e. two state or on/off) data and analog (i.e. values changing over time) data. Normally, the intent is to monitor and collect information from the field using PPCs, and then effect control over the processes using both automatic control programs as well as operator-initiated actions through workstations.

But consider how much information can be obtained from a single bit of data: the run status of a motor or pump. This field signal is discrete in nature, and is connected to a PPC from a dry contact closure. The status might come from the motor starter, where there is an auxiliary contact which closes when the motor is running; or the status might come from a flow sensing device downstream of the motor/pump which detects the presence or absence of flow. Either way, this single on/off status indictor can produce considerable data within the SCADA system.

At the most basic level, this 'running' status contact indicates whether or not the pump or motor is running. Typically the PPC turns on or off a discrete output which is wired to the motor starter circuit; a feedback signal, like the auxiliary run contact, is then used to confirm that the device is running. Within the PPC control program, timer logic can be used to verify that the motor starts and stops within an allowable time period. When the run control signal out is True, the logic monitors the feedback run status to determine if the motor actually starts; if the timer expires before the run status becomes True, then the motor has failed to start when commanded to run. Similarly, a stop control signal out can be compared against the false status of the motor run signal, to determine if the motor has failed to stop when commanded.

Using this run status input signal and some timers and counters, the PPC control program can accumulate the runtime of the pump or motor; as long as the run status input is True, then the timer and counter will accumulate the runtime for that pump or motor. Equipment runtime can be accumulated in hours, minutes, and seconds, or in hours and tenths of hours. Water and wastewater systems are programmed to maintain runtimes for all of the pumps in the system.

Next, a runtime limit or alarm limit could be programmed such that when the runtime exceeds that limit, an alarm is generated to the user at a workstation; this alarm indicates that the motor has operated for more than the allowed total time. A limit of 4000 hours, for example, might be used to indicate that the motor requires servicing. After the set time period, the alarm indicates that the motor should be taken out of service for maintenance; after servicing, the runtime can be reset through the PPC, and a new runtime accumulation can begin. Of course, the runtimes for all motors can be saved in the system for information purposes.

Suppose the programming has been set up so that runtimes are accumulated for each device for set periods, such as a week or a month; now the runtimes for that period of time can be entered into a spreadsheet for comparison purposes. Since spreadsheets allow for determining the maximum and minimum values, as well as average values, the spreadsheet can offer even more information besides how long each device has been operated in a set period of time. Comparisons can be made using bar charts and pie charts. Such comparisons might reveal if any motor has been operated longer than the other motors; in this case, a supervisor could investigate why the motor or pump has been operated either too long or for too little a time. Perhaps a pump has been underutilized, in which case its runtime will show as being lower than the others. Consequently, accumulating runtimes over time periods can reveal additional information about the operation of the equipment, in this case, the pumps or motors.

If the flow ratings of the various pumps is known, then a program could combine the flow ratings of the pumps with their respective runtimes, to derive a theoretical amount of flow for each pump. Since the actual measured flows are also being accumulated in the SCADA system, the program could then compare the calculated flows against the actual flows, thus identifying the efficiency of the pumps. A flow limit could be established such that any pump which is pumping less than the calculated flow by this limit could be 'flagged' for investigation. If a pump is operating significantly lower than the designed flow amount, then this condition can be determined by comparing the calculated flow with the actual flow, and using a limit or setpoint comparison. This then would provide one indicator as to the efficiency of the pumps in the system.

Imagine, all of this information can be derived from a simple discrete contact signal, one which is present on every motor and pump unit on every system! If this kind of analysis is possible for a pump or motor based upon its run status and calculated data within the PPC, imagine what other information could be derived by exploiting the data being collected by the SCADA system. The purpose of this example is to show that SCADA systems not only monitor field conditions for control purposes, but they can also accumulate information which can then be analysed to identify how efficiently the overall system is functioning.

3 Practical Procedures for SCADA Software Development

Objectives

- Describe the life cycle of a SCADA system project
- Identify procedures for developing the application software
- Understand how databases are created
- Explain how programming software should be developed
- Test and commission the application software

This chapter of the book presents an overall approach to designing and developing the software for a Supervisory Control and Data Acquisition (SCADA) system or simply methods and procedures which can be used to produce a well-designed and well-organized software system. While many system integrators and programmers may have 'their own way of doing things', the methods outlined herein are based upon many years of experience and have proven to be effective, compliant and appreciated.

Two of the key benefits of the methods described here are well-organized and structured software and well-detailed software documentation once the project has been completed. In addition, 'shop drawings' are produced as the software development progresses, a very important consideration when consultants are managing the project.

While this chapter strives to offer some guidelines in the design and creation of SCADA software, later chapters will address each of the topics here in more detail, including real world examples showing how these ideas have been used on projects.

3.1 Life Cycle of a SCADA Project

Before describing the process of designing and developing the application software for the typical SCADA system, an overview of the process involved in the creation and follow-through of a SCADA contract will be explained. The purpose of this section is to enlighten the reader on the various steps that a SCADA project goes through, and therefore what documentation is required and why.

Designing SCADA Application Software. DOI: http://dx.doi.org/10.1016/B978-0-12-417000-1.00003-8

3.1.1 Initial SCADA Project Specifications

The typical SCADA system project is part of a bigger process project, such as the design and construction of a water treatment plant or other process facility; the SCADA system is part of this overall project. In the case of implementing a SCADA system for an existing plant, then the SCADA project definition proceeds based upon the existing facility. In both cases, the consultant must first identify all of the processes involved and therefore identify the extent of the SCADA equipment (e.g. how many Programmable Process Controllers (PPCs), how many SOWs and how many signals, and so on).

The SCADA portion of the project, whether it is part of the bigger project or a SCADA system only, must be defined in terms of the equipment required. This includes the number of controllers, workstations, input and output field signals in each process area and any preferences that the customer may have. On this last item, a customer may already have a brand of field controllers and therefore wants to make the new system compatible. Contract documents must be prepared which include:

- Specifications for field controllers (PPCs) and operator workstations (SOWs)
- Identification of input and output signals required for each controller
- Process narratives for processes – initial version of Process Control Logic Descriptions (PCLDs)
- Process & Instrumentation Diagrams (P&IDs)
- Details on the construction of the works.

3.1.2 Development of SCADA Project

Once the project has been tendered and a contractor has been assigned, the contractor begins to design the system based upon the contract documents. The documents specify the shop drawings to be submitted, which invariably include expanded Process Narratives and Points Lists for each PPC in the system. Note that two of the software documents described in this book are the narratives (PCLDs) and the points lists (spreadsheets for each PPC). So if the contractor were to follow the guidelines presented in this book, then the required documents would be produced somewhat automatically.

Design submissions in the form of shop drawings for SCADA software consist of the expanded narratives as well as points lists for each PPC. The PCLDs become the expanded narratives. I have submitted my PCLDs for every project I was involved in, and the consultant and customer were very satisfied with the detail provided in these documents. Furthermore, once these documents have been reviewed and approved, any changes by the consultant will be considered an extra to the contract; so the contractor has protected himself against anyone adding 'a few little items' as it always seems to be. With the PCLDs, the design is well defined and established, and any change in how a system will function that is not described in the PCLDs would therefore be considered an extra to the contract.

During the development of the application software, there will usually be additions and changes to the signals to the PPCs and changes to how the system is intended to operate. Any change from the shop drawing submission, as outlined previously, constitutes a change in scope and is therefore subject to extras on the contract. The changes, regardless of how they occur, must still be incorporated into the software documents. As signals are added and/or deleted to the project, the points lists should be updated so as to avoid major editing to the documents at the end of the project. Also, by implementing these changes as the project proceeds, modifications to documentation are small and quite manageable. Likewise, the PCLDs should also be modified as changes are made for the same reason. Of course, any change in the contract should be approved before the changes are incorporated.

3.1.3 Factory Testing and Demonstration

Many projects require a Factory Acceptance Test (FAT) prior to going on site. This testing phase involves demonstrating to the consultant and to the customer that the specified design has been properly implemented in the software. Inputs are simulated to test the operation of the software. Since the software documents have been kept up to date, these same documents can be used to verify the operations of the software.

For example, inputs from either the field or the workstation software may be simulated to start one or more pumps. The software should then behave as per the documentation. Verification logic for testing that pumps start and stop within the allowed time can be checked by first turning on the 'run status' through simulation and then not turning them on to verify that the check logic is functioning correctly.

There are often changes identified during the FAT after seeing how the system functions in a simulated manner. These changes are again extra to the contract but must never the less be incorporated. Again by maintaining the documentation as the project proceeds, the contractor will have only small changes to make in the software documentation, rather than a large time-consuming effort at the end of the project; the end of projects is typically where the budget has been exhausted, and it is usually the documentation that suffers.

Getting through the FAT is another milestone in the project, as the software has been proven in design. The next step is to place the equipment in the field and verify everything again using the real world inputs and outputs.

3.1.4 Commissioning and Site Acceptance Testing

The Site Acceptance Test (SAT) is performed after all aspects of the system have been tested and verified. Before this final stage, the complete system must be tested and commissioned.

The first step in commissioning is usually to verify all of the physical inputs and outputs to each of the PPCs. A copy of the points list spreadsheets can be generated with a column for checking off the points as they are verified. For analog signals, this column could be used to document the scaled values which can later be incorporated into the software documentation. One of the issues that comes up at this stage is the sense of discrete inputs: are they Normally Open or Normally Closed type? This is one of the things that can sometimes differ between the original design and the field implemented system. The programmer can then make whatever changes are required and again mark up the documentation.

The commissioning proceeds one subsystem at a time. Like the FAT, the software must be tested against the PCLDs to verify that everything works as it should. The software documentation provides the specification for all of the software functionality. If some part of the software functions as per the design documents, but the consultant or customer claims that this should operate differently, the design must be reviewed to determine how the system should function. In some cases, the original documents prevail, while in other cases, a change is required. Once again, having the updated software design documents provides an approved specification for how the system is to perform.

Chapter 11 in this book describes a stepwise approach to test the complete system. For example, by having input and output processing subroutines (explained later in Chapter 9), one can simulate inputs and outputs without affecting or requiring the real world signals.

I have had situations in which the commissioning phase highlighted cases that were different from the documentation. Because I had been thorough in keeping track of all changes, and in keeping the documentation up to date, it was clear whether this difference constituted a change in scope or simply a misunderstanding in the design. I have had situations in which a change required was not worth a formal change, so I implemented the change for free (provided it did not cost me too much!). Some changes during this phase can result in goodwill and demonstrate to the consultant and customer that you, the programmer, are interested in giving the customer what he or she really needs, which could differ from the original design specification.

Ultimately, all of the subsystems will have been tested, adjusted and commissioned. One final step in many SCADA projects is the SAT. This is usually a continuous operation or run of the complete system for a period of 1–2 weeks without any major problems. During the course of regular operations, there may be conditions that arise quite simply, which no one thought of earlier. The parties must again meet to discuss how to handle the situation and decide if the current operation is correct or if a change is required. Again, any changes to the contract will be extra, and the software documentation will have already established a baseline or milestone in the design of the project.

3.1.5 The Final Step – Documentation

With everything having been tested and verified, the contractor only has to finalize the software documentation and submit it to the consultant. Stepwise updating will make this final step simple and require minimal effort. This final step may only require printing out all of the documentation for submission. All of the software documentation described in Chapter 4 addresses all of the requirements of any SCADA project.

In summary, it can be seen that starting with good software documentation, and maintaining it throughout the project, will save considerable time in the long run and eliminate the major updating at the end of the project, when the budget has typically already been exhausted.

With the overall project development in mind, now consider the detailed processes for designing and developing the application software for a SCADA project.

3.2 Identifying Process Area Field Signals

The typical SCADA system is designed to perform multiple operations in a number of areas within the overall system. A processing plant, for example, might have five or more defined areas, each of which is responsible for one aspect of the overall operation. A bottling plant might have process areas for queuing bottles into a line, filling the bottles, capping the bottles, grouping the bottles for packaging (e.g. cases) and palletizing the cases for shipment. Each of the process areas of the SCADA system therefore has specific monitoring and controlling requirements.

The water treatment plant, introduced in the last chapter, could have five process areas: raw water pumping, pretreatment, filtration, treated water pumping and chemical injection. Additional processes might include wastewater handling and the various remote pumping stations. Again, each of these process areas or areas of operation would typically involve a PPC with well-defined monitor and control functionality.

3.2.1 Process Areas and PPCs

The PPCs are normally installed inside control panels, where all of the electrical field signals are routed into. The location of the control panel is chosen to minimize the wire runs from and to the various field signals while making the panel accessible. The control panel, then, represents the 'heart' of each of the process areas of the total system.

Occasionally, there will be signals which are close to one PPC but which are actually needed by another PPC in another area; one solution to keep the wiring short is to wire these signals to the nearest control panel and then have the respective PPC transfer the field signals to and from another PPC in another panel using the interconnecting network. This network is typically a high-speed communication system using Ethernet or other communication protocol method.

Many projects involve a consulting engineering firm that issues contract documents for the complete project; the SCADA software development is one part of the entire project. These contract documents may already have the process areas defined, along with the required field signals to be included with each of the process area PPCs. It is therefore up to the software developer, the system integrator, to ensure that the software is designed and developed in accordance with these contract documents.

3.2.2 Identifying Input and Output Signals

Each process area would require a PPC to monitor and control the operations within that section of the system. Having identified the specific areas, the next step is to define and identify all of the field inputs and outputs associated with each area.

This requires considering what field equipment is involved and how each device is intended to be used and/or operated. A pump, for example, would have discrete signals for run status and run control and perhaps analog signals for discharge valve position and control. There would also be area signals such as reservoir or tank levels, analog signals to monitor and general alarm conditions.

Whether the consultant has already prepared the list of field signals to be included with each process area, or the system integrator is responsible for developing it, the list of all signals to be included must be organized and tabulated. As will be described shortly, the spreadsheet offers the perfect method of organizing and generating these points lists.

In developing a points list for a process area, there are two general categories into which the field signals fall: equipment-specific and general process signals. These signals need to be tabulated and organized into groups, such as all signals for each pump or valve. Some signals relate to the whole process area and not to any one field device. In the end, a comprehensive list of all field signals must be prepared.

The equipment-specific signals are those that connect to a unique device, such as a motor, pump or conveyor. In the water and wastewater industries, consider the field signals associated with a single pump, organized by the four basic signal types:

Discrete Inputs	Pump Run Status
	Pump General Fault or Alarm
	Discharge Valve Open Status
	Discharge Valve Close Status
	Local/Remote Selector Status
Discrete Outputs	Pump Run Control
	Discharge Valve Open Control
	Discharge Valve Close Control
Analog Inputs	Discharge Valve Position Feedback
	Pump VFD Speed Feedback
Analog Outputs	Discharge Valve Position Control
	Pump VFD Speed Control

The general process area signals are those pertaining to the process area as a whole, such as a well or reservoir level or a flowrate from the location. Using the water application, some general signals, organized by signal type, might be as follows:

Discrete Inputs	Raw Water Well High Level Alarm
	Raw Water Well Low Level Alarm
Discrete Outputs	Audible Alarm Annunciator Silence Control
Analog Inputs	Raw Water Well Level
	Raw Water Discharge Flowrate
	Raw Water Discharge Valve Position Feedback
Analog Outputs	Raw Water Discharge Valve Position Control

In Chapter 6 on Developing the Application Program Databases, it will be shown how to organize all of these signals in a logical manner using spreadsheets. The I/O modules for the PPC would be selected in such a manner that the signals can be grouped in a logical way. For example, all of the device-specific discrete input signals would be arranged on the same I/O module, while the more general system signals might be assigned to another I/O module. Each of the modules in the PPC chassis would then have related signals together, making it simpler to locate specific signals in the system.

3.2.3 Organizing Field I/O Signals

Once a complete list of all field signals has been tabulated for a given process area, the next step is to determine the number of I/O modules required for the PPC and how the signals are to be assigned to the modules. Even if a consultant has specified the field signals to be included in each process area, the various signals must still be assigned to individual I/O modules for the configuration of the PPC.

Generally, it is good practice to group signals from one piece of equipment (e.g. pump) together on one module; in the unlikely event of a module failure, only that one piece of equipment is affected. An alternate approach to assigning signals is to spread the equipment signals over multiple modules so that a module failure only affects some signal loss for equipment. Unfortunately, these signals losses become spread over more than one piece of equipment. It is the author's opinion that grouping signals for each device into one module each is the preferred approach.

The general process area signals, such as the good level and the analog monitor signals, can be assigned across the modules in the PPC. Where there are multiple analog input signals, for example, it is good practice to assign these signals over more than one module, so that again, a module failure will not result in the loss of all the process signals.

3.2.4 Allowing for Future Expansion

It is generally understood in the SCADA systems industry that for the PPC, there should typically be 20% spare capacity of both points on a module and spare slots or spaces in the rack. A discrete input module might have 16 total points available, of which 13 are assigned; this leaves 20% capacity on that module. Alternatively, additional discrete input modules can be included to which few if any signals are assigned. Every SCADA system gets expanded at some point in time, so having this extra capacity for future expansion represents good foresight.

There should also be spare slots in the PPC; whether the PPC consists of a single rack, or many racks interconnected, there should be spare capacity in the available slots also. The same 20% rule applies to spare slots also. As an example, if a PPC has a total of 97 field signals wired into 10 modules, then there should be an additional 20 spare points spread over the 10 modules. And for the 10 modules, which use 10 slots, there should be perhaps two empty slots in the rack(s).

3.3 Creating and Documenting Application Databases

SCADA software requires multiple databases: each PPC has a database for field and program points; the SOW has a database for the graphic display information; and the tagname definitions form a database to be used for creating tagnames. Before a database can be created, there should be some standard form or structure to the tagnames to be used in the databases. Then each of the points in the various application programs can be assigned tagnames based upon the structured naming convention developed.

3.3.1 Developing a Standardized Tagname System

Every program point in a program requires a unique name, called a tag or tagname. Whether a point represents a real world field signal or an internal or virtual program point, every point or signal must be uniquely identified with a tagname. As will be explained in a later chapter, standardizing on the method of naming signals and points simplify the tagging system as well as create a structure to the tagnames; both current and future points can be assigned tagnames based upon a well understood system.

To illustrate the concept of a standardized tagname system, consider some signals associated with a pump; all of the signals listed here relate to the same piece of equipment, and the tagnames identify the specific signal or data point associated with that pump:

LLF_RWP2_DV_SOD Raw Water Pump 2 Discharge Valve Open Status
LLF_RWP2_PB_YST Raw Water Pump 2 Host Start Pushbutton
LLF_RWP2_DV_QZI Raw Water Pump 2 Discharge Valve Position Feedback

Comparing the tagnames shown here, it can be seen that they all share the first two parts or fragments: the LLF for Low Lift and the RWP2 for Raw Water Pump 2. The remaining fragments further define the signal, using additional short forms or abbreviations. If a new signal was to be added for this pump, then the first two fragments would be the same and the remaining parts would identify the specific signal.

By structuring the tagnames for a SCADA system, the assignment of a tagname simply requires picking phrases or fragments from 'dictionaries' that are created during the design phase. These dictionaries are simply lists of the valid fragments for each part of a tagname.

3.3.2 Creating Points Lists Using Spreadsheets

Since the points lists for the PPCs and the SOWs need to be organized and recorded, one could use a spreadsheet program (i.e. MS Excel). Using the spreadsheet program, multiple sheets can be created and organized to contain all of the points organized under different headings, such as tagname, description, address on PPC module, panel terminal numbers and point type. The finished database product would consist of several sheets of spreadsheets identifying all field and program points used in each of the parts of the SCADA system.

This method of using the spreadsheet to create points lists is addressed in detail in a later chapter; for now, consider the basic concepts of using spreadsheets for tabulating all of the points being used in the application software. A spreadsheet template can be created with the required headings on the columns; this template sheet can then be copied any number of times to accommodate the number of I/O modules in the PPC chassis. The suggested headings include:

- Tagname and Description
- Point Type – DI, DO, AI, AO

- Address on Module
- Module Code
- Terminal Numbers

3.3.3 A Spreadsheet Illustration

In Figure 3.1, a sample spreadsheet is shown, with the signals organized and assigned to modules in the PPC. Each signal is given a tagname, which follows the structured tagging system, plus a description of the signal, the type of field I/O point, the address of the signal within the PPC and the module catalogue number. As will be explained in more detail in Chapter 6, these spreadsheets can be created for all PPCs in the complete system.

3.4 Defining and Documenting the Process Controller Operations

There is always a need to fully document the detailed operations of each of the process areas in terms of what control functions are being performed and what key signals are involved. A detailed description of each of the process operations in each of the process areas, from the perspective of the program logic, must be prepared such that the reader can understand how the application software implements the required monitor and control operations. The document for this is referred to as the PCLDs. This topic is addressed in detail in Chapter 7.

3.4.1 The Initial Process Narrative

In the initial design of a SCADA system, the processes must be identified by area or part of the overall system. The overall project must be divided into process areas, which may be physically separate as is often the case in a water treatment or wastewater treatment plant. The initial process narrative should consist of the following basic elements:

System overview

This description provides a summary of the monitor and control functions of the application. The specific process areas are identified along with the process controllers (i.e. PPCs) that are required.

Application architecture

An initial architecture figure should be developed that illustrates how the previously identified components are to be interconnected. Using the water treatment plant as an example, the initial architecture might appear as shown below in Figure 3.2.

I/O address assignments

As described in Section 3.1, the various field I/O signals for each of the process areas must be tabulated or listed, by controller; this may change during the design and development of the project, but these initial points lists provide a starting point for defining all of the process operations to be included.

Water Treatment Plant **Raw Water Pumping Station**		*Controller Name:* RWPSTN *Network IP:* 192.168.205.141			
Tagname	Point Description	Point Type	Hardware Address	PLC Module Prod. Code	Terminal Numbers
LLF_RWP1_00_SRN	*LL Pump 1 Running Status*	DI	I:7/0	1746-IA8	
LLF_RWP1_00_SCM	*LL Pump 1 Remote Mode*	DI	I:7/1	1746-IA8	
LLF_RWP1_00_AGF	*LL Pump 1 General Fault*	DI	I:7/2	1746-IA8	
LLF_RWP1_PS_APL	*LL Pump 1 Low Pressure Alarm*	DI	I:7/3	1746-IA8	
		DI	I:7/4	1746-IA8	
		DI	I:7/5	1746-IA8	
		DI	I:7/6	1746-IA8	
		DI	I:7/7	1746-IA8	
LLF_RWP1_DM_DRN	*LL Pump 1 Motor Run Control*	DO	O:4/0	1746-OA8	
LLF_RWP1_DV_DON	*LL Pump 1 Disch. Valve Open*	DO	O:4/1	1746-OA8	
LLF_RWP1_DV_DCE	*LL Pump 1 Disch. Valve Close*	DO	O:4/2	1746-OA8	
		DO	O:4/3	1746-OA8	
		DO	O:4/4	1746-OA8	
		DO	O:4/5	1746-OA8	
		DO	O:4/6	1746-OA8	
		DO	O:4/7	1746-OA8	

Figure 3.1 Sample field I/O points list spreadsheet.

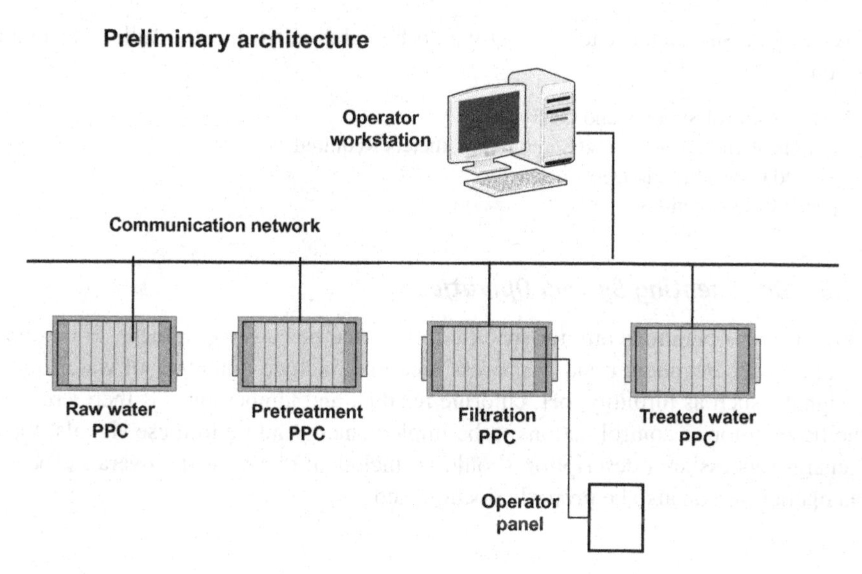

Figure 3.2 Initial system architecture.

For each such process area, the specific process operations must be identified and described. The initial process narrative might describe how the pumps are to be operated, such as duty assignments and control setpoints. General concepts about the process operations can be written, with the details to be filled in later. But the general operating concepts must be established.

3.4.2 Developing the Detailed Process Logic Descriptions

With the process areas and the individual operations identified, the next step is to describe the details of each process area. These descriptions would include the equipment involved, the setpoints or parameters used in the control operations and any special considerations such as hardware and software failure handling. The user can understand from these descriptions exactly how the application program operates to effect the operations described.

Now the details of the operation must be defined and explained. What control setpoints are required? How should failures and alarms be handled in this area? Will there be duty assignments for the equipment? What modes of operation will be available, such as Remote Manual and Remote Automatic? It would certainly help to have a standardized format for these detailed descriptions, and this is what is described later in Chapter 7. A single process area may have more than one logic description, and therefore there should be separate logic descriptions for each of the processes. These descriptions can then be organized into chapters or sections of the PCLDs document.

As will be seen later, each PCLD would be subdivided into the following main sections:

- System Control Strategy and Overview
- Equipment and Facilities and Setpoint Parameters Required
- Detailed Control Logic Descriptions
- Special Considerations

3.4.3 Documenting System Operations

Some process operations are not specific to any one piece of equipment. For example, in a water treatment plant, a process area may include a number of water quality signals, such as turbidity, pH, chlorine residual and temperature. If there are any specific monitor or control actions to be implemented relating to these signals, then a general process area description should be included; this way, the overall process area operations can also be properly documented.

3.5 Designing and Developing the SCADA Application Software

With the field I/O points lists and the detailed logic descriptions for each PPC in the project defined, the application software or programming can be started. Since both the PPC and the SOW act on information in the SCADA system, it is necessary to prepare the detailed points lists and process operations descriptions for all areas of the system.

3.5.1 Developing the Controller (PPC) Application Software

The program logic for each controller can be developed from the PCLDs and the field I/O points lists. As the programs are evolving, additional software database points will be defined and created for each PPC. Hence, the design and development of the PPC application programs and the refinement of the database will require an iterative process.

In the past, the PPC programming software supported only one program, which ran continuously. All operations had to be implemented in subroutines within this program. Today's PPC programming software, such as A-B RSLogix 5000, allows for multiple programs in multiple tasks. The basic continuous or base task with programs can still be used for the bulk of the programming, but some additional features allow for special cases.

Separate tasks can be created that execute at regular intervals, such as every 50 ms; cyclic operations such as PID control logic can be programmed into one of these synchronous or cyclic tasks to ensure that the logic executes precisely when it is needed to. Another use for these cyclic tasks is data collection, such as production totals by the hour or day or water quality data from a water treatment plant.

The programming software also provides for event or asynchronous tasks, which execute when a particular condition is present. For example, separate fault handling logic can be created for each of several abnormal or alarm conditions that may occur. Alternatively, a single such event task might handle a group or category of event-based situations.

With these additional structures in the program logic, the programmer must consider how best to organize the program logic into functional sections; that is, how many tasks and how many programs within the tasks are required to implement the logic defined in the process logic descriptions.

Every possible operation that could be enacted in the controller must be allowed for. As an example, if the modes of control include Manual and Automatic, then the program logic must include interfaces between the SOW graphic displays (e.g. start and stop virtual buttons) and the PPC logic; for automatic control, the logic or control algorithm must be programmed to allow for accessing the setpoints that the user may enter via the workstations.

3.5.2 Developing the SCADA Workstation (SOW) Application Software

The application software for the SCADA Workstations (SOWs) consists of two major parts: the process graphic displays which present information to the user and the databases containing all of the information from the PPCs. The process graphic displays include current information such as status of equipment and current values of field signals. Other displays can be created to provide operator input for entering setpoints and changing equipment duties.

In order to present this information, there are typically two databases maintained in the workstation application: current or process and historical. The current database contains all of the values and information for every point configured in every controller. The communication driver or server collects data from the PPCs at regular intervals, thus keeping the current database up to date. The historical database maintains values over time, so that historical trends and reports can be generated from the collected data.

The application software for the user interface consists of various current and historical displays. These must be defined and organized around the information being collected in the databases. Issues such as colours to be used, navigation methods among displays, and the hierarchy of the displays must be considered when designing the process graphic displays.

Security is another issue to consider when designing the SOW software; what levels of access are to be allowed and how a given user 'logs in' to the system. Operators can access any of the process displays and view the trend displays. But displays containing setpoints and parameter adjustment might be limited to supervisory personnel. And making changes to the system configuration and application software is normally restricted to the programming personnel, those who actually developed and/or maintain the complete SCADA system.

Hence, the application software for the workstations involves the development of the databases, the process graphic displays, the development of a navigation methodology and the creation of historical displays, to name a few. The starting point, however, should be the development of the current database, as this is built up from the individual PPC databases.

3.5.3 Networking and Interfacing

To allow the workstations to communicate with all of the controllers in the system, a communication network must be selected and designed. This typically requires the selection of appropriate software drivers to access the controllers over the selected network system. The most common network system uses Ethernet with TCP/IP, but there may be special devices requiring other communication interfaces. For example, a Profibus interface may be required if one or more controllers are Siemens PLCs whose standard interface is Profibus. It may be possible to use the same communication protocol for all controllers, but this must be taken into consideration when designing the network and the method of communications.

3.6 System Integration and Checkout

Once all of the application software has been developed, it is time to test it. An organized approach is needed to commissioning, since there will be many processes to be tested with multiple controllers. This topic will be covered in detail in a later chapter, but for now, consider some basic ideas.

3.6.1 Controller I/O Signal Checkout

With PPC field I/O points lists in hand, each of the signals to and from each controller can be verified. Before any commissioning of the software can begin, it is important to know that the physical signals are operational and wired to the correct inputs and outputs. Once the I/O checkout has been 'signed off', then the application testing can begin.

3.6.2 Testing and Commissioning Process Area Operations

From the PCLDs, the programmer can isolate specific operations and test each one separately. A procedure should be used whereby an electrician or a plant supervisor can guide the programmer through the individual tests. Since some parts of the system may not be completed, the project manager should direct the commissioning to those areas which have been completed and are therefore ready.

3.6.3 System Acceptance Test Procedures

Some projects may require someone to prepare detailed procedures on how each part of the system is to be tested. Such procedures may be general guidelines or very rigid steps for testing. Either way, a methodical approach should be adopted to commissioning since most systems consist of several process areas and operations.

Some projects involve the creation of detailed test procedures for both the factory and the site. Factory testing is sometimes required to confirm that the programming has been developed in accordance with the contract specifications. This testing is done at the contractor's facility, prior to delivering the equipment to the project site. The intent of this testing is to confirm that the program logic functions as it was intended, using simulation of signals to verify the program logic. The site testing consists first of the field I/O testing, followed by a system by system checkout of all aspects of the program logic. After this site testing, the project is considered substantially complete.

3.7 User Operations Reference Manual

This document is intended for the user of the SCADA system. The system is viewed from the perspective of the operations workstations through the process graphic displays and reporting systems. All of the information being presented along with each of the control operations available to the user must be documented for the end-users of the system.

3.7.1 Purpose of the User Operations Manual

The PCLDs described earlier are intended to provide a detailed design explanation of the operations of each of the application programs in the PPCs. The user manual, in contrast, provides an explanation of what the user can do through the workstation and how it can be done. This document assumes that the reader (user) has no knowledge of programming and SCADA systems but does know how the facility is intended to operate. Consequently, the user manual must be written for this audience.

Aspects of the SCADA system which should be incorporated into this manual include: general description of the system, security and passwords, alarm and event processing, specific operations available to the user and historical trends and reports which can be accessed.

3.7.2 Structure of the User Operations Manual

SCADA systems vary significantly in the complexity and number of processes being handled. A simple system could be made up with as little as one PPC and one SOW. Regardless of the size and complexity, however, there needs to be a detailed description that covers all operations of the system. A starting point for the organization of the user reference manual could include the following major sections:

- SCADA System Overview
- Process Graphic Displays
- Alarm and Event Processing
- Historical Trend Displays and Reports
- System Maintenance Procedures

In a later chapter on Documenting the SCADA Operations, more detail will be provided on the content and organization of this user-oriented document.

4 Documentation for SCADA Systems

Objectives

- Define and describe the major documents for SCADA application software
- Describe the structure and content of the database document
- Explain the purpose of each of the documents

For any large project, contract documents are first prepared, reviewed by various parties and then tendered for contractors to bid on. Once the project has been awarded, the contractor designs and implements the designated project in accordance with the contract documents. As the project progresses, there are often changes made, which must be reflected in the contract documents. At the final submission stage, updated or 'As-Built' contract documents must be prepared and submitted back ultimately to the customer for record purposes. These documents then serve as a record of exactly how the entire project has been implemented.

Supervisory Control and Data Acquisition (SCADA) software should follow the same approach, although all too often, the documentation consists only of Programmable Process Controller (PPC) program listings which may or may not include some comments. There is often no narrative description that describes all of the details as to how the PPC program does what it does. Likewise, there is no record of how each of the process graphic displays has been constructed and animated; the process database which is critical to the operation of the SCADA Operations Workstations (SOWs) is not documented either. It would be most useful to have a document which summarizes all of the hardware and software signals used in the project, including details such as which PPC the signals apply to, the type of data they represent (i.e. boolean flag and analog input real) and the details of each tagname used.

4.1 SCADA Software Documentation

This chapter presents an approach to develop a complete set of documents for any SCADA software project, regardless of size and complexity and of application. Before presenting the components of software documentation, it is worth considering the benefits or reasons for developing this level of documentation.

Designing SCADA Application Software. DOI: http://dx.doi.org/10.1016/B978-0-12-417000-1.00004-X

4.1.1 Reasons for Developing Documentation

In addition to what was presented in Chapter 1 on the need for software documentation, consider the following reasons for having such documentation; with proper software documentation, this information:

- aids in the assignment of hardware and software addresses,
- aids in the installation work during startup, as all addresses have been identified and located within the PPC chassis,
- helps with troubleshooting, especially during commissioning,
- explains how the system was designed to operate,
- identifies spare inputs and outputs on the PPC for expansion,
- offers a lay person's explanation of how all of the parts of the system function, including the process control operations with their respective parameters,
- illustrates the structure of tagname such that understanding the PPC program logic becomes easier to follow.

4.1.2 Components of Software Documentation

As will be seen, the software documentation can be organized into five basic components, which are briefly described below:

Database Reference
Explanation of the tagging system used and the spreadsheet points lists for all field I/O and all internal program points used in the various application software; points lists are provided for both the controllers and the Human–Machine Interface (HMI) workstations.

Process Control Logic Descriptions
Detailed description of the operations of all PPC application programs, identifying the key signals and parameters used and the detailed explanations of the program logic.

SCADA User Operations Reference
Explanations of each of the displays and reports available to the user, including all control modes available and the procedures for effecting control of the field equipment, along with system maintenance procedures.

PPC Application Program Listings
Summary description and program logic listings for each controller in the system, include controller configuration and settings.

SOW Application Software Reference
Configuration and organization of the displays, databases, scripts and I/O drivers used in the SCADA application.

4.2 Database Reference

The database reference document contains all of the lists of program points used in the PPC and the SOW; this includes both field input/output signals as well as the internal or virtual program points or tags created within the program logic. These points lists are created in spreadsheet form, as they can be easily imported into both the PPC and the SOW application software. This import operation will be addressed in more detail in a later chapter of the book.

4.2.1 Standardizing on Tagname Conventions

The description of the Tagname Signal Naming Convention (TSNC) should be incorporated into this document, along with the various tagname fragment or element dictionaries (refer to later section for details). In a following section, the TSNC details will be provided, but basically each point will be composed of a structured tagname which includes:

- Area or location such as pretreatment and remote station
- Equipment code, such as pump, valve or device
- Equipment component code, such as discharge valve or transmitter
- Signal type such as analog, discrete and virtual
- Signal designation such as open status, run/stop status and analog value.

To illustrate this concept of standardized tagnames, consider some examples showing the interpretation or explanation of the tagnames. Following are some tagnames using the dictionaries and structure described herein. It is hoped that the methodology used will become clear, and that the value of such a system will also become clear.

```
HLF_TWP3_DV_SCD
    Treated Water Pump #3 within the High Lift Station
    Discharge Valve closed status input
FLT_FG07_FT_QFI
    Gravity Filter #7 of the Filtration area
    Flowrate transmitter for filter, current flowrate
PRE_SIT4_IV_YON
    Silica Tank #4 of the Pretreatment facility
    Inlet Valve to the filter; virtual open command from program
```

4.2.2 Programmable Controller Database

For the PPCs there should be two sets of spreadsheet listings: one set lists all of the physical or hardware input and output signals connected to the PPC and the other set lists all of the internal program points used in the application program. As will be shown later in this book, the creation of and formatting of these spreadsheets can include all of the information about the hardware and software points used.

For each PPC, which in turn represents a process area within the overall SCADA system, there should be a spreadsheet points list showing all of the points used for the process area. Such spreadsheets would be organized by I/O module, showing the details of each signal connected to each point on each module. Since all application programs have internal or software points as well as the I/O points, another set of spreadsheets would list these points with their attributes, including the tagname, the nature of the point (e.g. boolean, real and integer) and optional description.

The controller database therefore consists of two major parts:

- Hardware: Field I/O Signal Points Lists
- Software: Virtual Program Points Lists

It should be noted that the virtual points actually consist of two types: software program points and programming support points. The software points are those for which there is no physical address, such as a virtual flag that is set or cleared based upon several physical conditions and an analog setpoint entered via the SOW. A number of physical signals like general alarm, high temperature alarm and emergency stop might be combined to set or clear a virtual point called 'Motor Available' or 'Motor Ready'. These software points are used extensively in programming.

The programming support points are data structures used to enable some logic operations. Timers, for example, are used to verify that certain operations complete successfully within an allowed time; these timers would be considered support type points since they exist as points in the database but are not actually representing a particular signal. Timers, counters and arrays are some examples of these support type database points and should be included in the virtual program points lists.

4.2.3 SCADA Workstation Database

The SOW database will consist of the PPC hardware and software points, along with any internal virtual points used in the HMI application software. For example, a discrete input point from a PPC might include the following attributes:

- Tagname and Description
- I/O Driver Access Name and Address
- Text for On and Off states
- Logging and Alarming Characteristics

An analog input point from a PPC might include the following details:

- Tagname and Description
- I/O Driver Access Name and Address
- Scaling for Analog Raw and Engineering Values
- Alarm Setpoints and Deadband Values
- Historical Logging Settings

In summary, the Database Reference documentation provides both a detailed explanation of the tagging system used and detailed spreadsheet listings for every data point used in the application, including both the controllers and the workstations.

4.3 Process Control Logic Descriptions

Once the PPCs have been identified and their respective functionality has been defined, then detailed narratives, called Process Control Logic Descriptions (PCLDs), are required which describe in complete detail all aspects of the program logic. This document is organized by functional area. There would be four (4) major sections for the PCLDs, as summarized below.

4.3.1 Control Strategy Overview

The System Control Strategy Overview describes the 'big picture' of the application software, such as the purpose of this part of the system and the main monitor and control operations. The overview would describe the conceptual operation of the system, without any of the specific details of operation; the details are covered in a later section.

For example, the filtration system of a water treatment plant would provide a description of the filter operation in terms of flowrate control and backwash requirements. A summary of the operational concepts would be included in this section.

4.3.2 Equipment and Parameters

The Facilities and Parameters details the equipment involved, the key process signals involved in the process and the control setpoints or parameters used in the control. Again, at this level, the information is general but detailed in identifying the specific equipment used and the parameters that are used in the process.

The major pieces of equipment involved (e.g. pumps, valves and conveyors) would be identified to provide an introduction to the logic descriptions which follow. Any setpoints or parameters involved would also be identified, so that the configurable aspects of this system can readily be defined.

4.3.3 Control Logic Descriptions

The Control Logic Description offers all of the detailed information about the process or processes. If a system (i.e. PPC) handles more than one process, then separate logic descriptions should be provided for each such process. The detailed description would make reference to the tagnames used in the program logic, so there is no misunderstanding on how the program logic implements the intended monitor and control functions.

For the filtration system, the description would explain how the filters operate in terms of the signals used, the PID control parameters involved and the various field and virtual program points used in the program. This description might be organized into modes of control, such as local, remote manual and remote automatic. Each operating mode would be described in detail such that the reader would understand exactly how the application program implements the design intentions. For the filtration system, there would be two main processes to describe: the normal filtration control using a PID control algorithm and the sequential steps of backwashing a filter.

4.3.4 Special Considerations

Finally, this section would provide the 'what if' information or how abnormal conditions such as alarms are to be handled. In some systems, alarms may be directed to a telephone autodialer system for after hours; other alarm conditions may cause the program to shut down or disable parts of the process. Any data interfacing between this PPC and other PPCs would be identified, as data is often transferred between PPCs via the SCADA network.

If there are hardwired backup systems, these would be described so that the interaction of the software and the backup hardware can be understood. There are some instances in which safety or other considerations call for a hardwired backup, even though the application software would normally handle all operations.

4.4 Controller Application Program Listings

For each of the PPCs in the system, a complete detailed hardcopy report of the program logic and configuration should be generated. With today's controller software, like A-B RSLogix 5000, very complex program structures can be created involving multiple programs and multiple tasks. So in addition to the actual program logic listings, there should be a record of the programs, routines, tasks and I/O configuration for each PPC. This information would include the following details:

- Summary description of the program structure
- Programmable Controller configuration and settings
- Input & Output modules used and their location within the chassis(s)
- Program logic listing by routine.

The program listings serve as the main document for understanding and documenting the system operations; the other documents provide detailed support information to complete the understanding. Rung comments, descriptions for operands and other documenting features should be used in the programming software for the application programs.

4.5 SCADA Workstation Application Software Reference

The SCADA User Operations Workstations include process graphic displays, historical and current process databases, alarm and event summary displays, security and log-in displays, I/O driver configuration and settings and background operations which are often implemented as scripts. This document should provide the design details of all of these aspects of the application software.

The SCADA software executing on the workstations consists of two basic parts: the HMI facility which generates and updates all of the displays and reports and the I/O Server or Driver which provides communications between the workstations and the various PPCs in the system. The latter component executes in the background and requires no user intervention, except in the event of communication problems.

The communication drivers used to interface with the PPCs to retrieve data and issue commands must be configured with specific 'paths' from the SOW to the individual PPCs. For the WonderWare InTouch HMI SCADA software, the interface between the HMI software and the I/O driver software is defined as an 'Access Name', which consists of the following elements:

- Access name — typically the name of the PPC on the network
- Node name — name of computer on network performing communications
- Application name — name of the I/O driver or server program
- Topic name — name used in the driver/server

Each database point in the HMI application that references a PPC data point requires a reference to one of these Access Names. The I/O Server program is configured with details which provide a path from the workstation through the network to the processor of the PPC:

- Communication port on the host computer
- Communication module in the PPC through which it communicates
- Backplane or chassis connection
- Processor name in chassis.

There should be a display hierarchy and screen navigation methodology. Typically, there is an overview display from which the user can access all of the other displays. For example, the overview display shows the complete system in the form of blocks or key information; for water systems, perhaps the levels and flowrates for each of the system areas would be shown. Using a pull-down or pop-up menu of displays, the user would then navigate through to other displays. Clicking on the water treatment plant in the overview would show a more detailed display of all the key parameters throughout the plant. Clicking on the filtration area would then show a detailed display of the filters with all of their respective information.

Thus, a listing of all of the displays, regardless of type, should be produced with a navigational hierarchy showing how the user would access each of the displays; such a display would serve as a main menu for the user.

Another feature of most SCADA software is the scripting facility, in which simple programs can be developed to perform operations 'behind the scene' or in the background, concurrent with the operation of the user interface and the I/O server communications. As an example, pressing the key sequence 'Control-A' might be configured to display the Alarms Summary screen, regardless of which display the user is currently viewing. Another example might clear out data entry tags when the pop-up window is displayed, so that the user can enter new values.

These scripts should be documented and described, as they perform key operations in the background while the system is running. Typically a listing of each script should be provided.

4.6 SCADA User Operations Reference

The user of the SCADA system typically operates through a process graphics-based workstation. Every system and subsystem are represented by a series of colour-animated process graphic displays. A method of selecting displays and navigating through the displays is required. The user needs to know how to effect equipment control, modify setpoints and parameters, view current status of each subsystem and access the historical and reporting features of the system.

The SCADA User Operations Reference document provides a detailed description or 'how to' description for every display and every operation available to the user. Colour screen shots may be included so that the document can refer to the images, such as the 'Start' or 'Control' pushbuttons on the display. Any error messages or faults that might occur must be described so that the user can resolve the problem. For example, in attempting to start or stop a piece of equipment, the user may see a 'Failed To Start' message; the document should explain what conditions may have caused this message, such as the motor did not turn on within an allowed time period.

This user reference manual with its organization and structure will be covered in detail in Chapter 8. This document should be divided into major sections, each of which explains and illustrates a particular set of displays and operations. For this chapter of the book, it will suffice here to highlight the main elements or sections required in this system user reference document.

A suggested organization for this document would have the information divided into the following major sections; each is briefly described below:

- General System Overview
- System Graphic Displays
- Process Graphic Displays
- Historical Reports and Trend Displays
- Special Operating Procedures

4.6.1 General System Overview

This first section is intended to introduce the user to the concepts and general organization of the SCADA system. The system architecture and the overall arrangement of equipment would be illustrated and explained, so that the user has an understanding of how the complete system has been put together. The central control facilities, that is the SOWs, are described in terms of the role they play in the system and how they differ from the I/O server workstation.

The display conventions and navigation methods are described, as this applies to all of the displays in the system. Colour conventions such as running is red or green, alarm indication is yellow or a flashing colour and an event is blue; the choice of colours is important so that the user becomes accustomed to both the displays and what each colour means.

The design of the graphic displays may include a standard set of pushbuttons which appear on all displays; such buttons include the alarm summary, the system overview, the trend display menu and so on. This type of 'quick link' panel of pushbuttons can be very useful for accessing the more commonly used displays from any other display. In Chapter 10, there is an explanation of this type of panel and how it can be used.

Finally, the security levels and password setup are described; different operations may require different access levels. Starting and stopping equipment and changing the control mode for equipment are typically an operator-level function. Changing system parameters and operating setpoints may be a supervisor-level function. Full access would only be granted to programming personnel, who are responsible for maintaining the application software and the system as a whole.

4.6.2 System Graphic Displays

In every SCADA system, there are displays which present overall or system type information: a system overview for a water treatment plant would show the status of all pumps, the levels in all tanks and reservoirs and key process signals like flowrates and water quality. A main menu display would include links to the other detailed process graphic displays.

Often the system overview display includes hyperlinks to access the detail displays; clicking on the high lift area of the overview would show a detailed display of the pumps and other equipment in the high lift station. There may be other such global links that can be accessed from overview type displays.

4.6.3 Process Graphic Displays

Every SCADA system uses colour graphic displays to present current operating information; this section of the document would explain each of the process displays in terms of what information is being presented. The colour conventions used for display and the method of navigating through the various displays would also be explained.

Each process area of the SCADA system would typically have its own group of displays; hence, the user manual should be organized by process area, so that all of the displays for each area can be found in that section of the user manual.

The process graphic displays contain all of the details needed by operations staff to identify the current operating conditions; what equipment is running and what equipment is stopped; what the key process signal values are at any time; what mode of control the equipment is currently set for and so on. There may be additional displays and/or pop-up type displays which provide the user with additional detailed information.

4.6.4 Historical Reports and Trend Displays

When the trend displays are initially created, the default display characteristics are defined, such as the time frame for the *x*-axis, the points assigned to the trend and what colours are to be used for each signal in the trend. The historical trend displays can typically be configured while viewing them, so this section describes all of the trend displays available and how to configure them for specific signals and time frames.

There may also be historical reports, such as equipment runtimes and total flows or productions, which can be accessed. In some systems, this historical information can be exported for use by management personnel. Such reports are useful for maintenance purposes to determine when equipment is due for servicing, based upon the runtimes shown in the report.

4.6.5 Special Operating Procedures

Standard procedures should be defined for starting up the system and shutting down the system. There may be occasions when the central facilities need to be 're-booted' to correct a problem or to implement some application software changes to the system. Since shutting down the system is expected to be a rare event, having a detailed procedure in place in this manual will ensure that the process is carried out properly.

5 Tagnames and Signal Naming Conventions

Objectives

- Explain the concept of a structured tagname system
- Describe the components of a tagname
- Explain the purpose of each of the components of a tagname

Every hardware input or output signal requires a unique name or an identifier; every internal software point also requires a unique name or an identifier. These identifiers, or tagnames as they are usually called, are typically alphanumeric strings of characters, organized into a meaningful name. A common naming convention for all tagnames, regardless of where they exist, would provide not only a standardized tagging system, but also a tagname system in which the tagname itself identifies the location and characteristics of each signal.

5.1 Original Signal Tagnames

The Instrument Society of America (ISA) has had a tagging system in place for many years; this tagname system was developed to refer to signals used in instrumentation and control systems. As such, the system has been very effective and useful and is still in use today in the traditional Process and Instrumentation Diagrams (P&IDs). This system uses four letters, four digits, separated by a hyphen.

This ISA structure of four letters and four digits, in the form of AAAA-NNNN, could be considered the first structured tagname system. Figure 5.1 shows the letter assignments for the various signal types and how each letter is used in an ISA tagname. An ISA tagname would then be constructed using the letters from this chart, in the order indicated.

For example, the tagname FIT-201 refers to a Flow Indicating Transmitter whose reference number is 201. Another example is LIC-402 that refers to a Level Indicating Controller number 402. The ISA system is very extensive and covers virtually all signals used in systems today. However, this system does not identify where the instrument is located or where it is used. One must refer back to contract drawings and P&IDs to cross reference the tag and determine what it is referring to. Add to this the fact that software points in programs also need tagnames, the ISA

Designing SCADA Application Software. DOI: http://dx.doi.org/10.1016/B978-0-12-417000-1.00005-1

	First letters		Succeeding letters		
	Measured/Initisting variable	Variable modifier	Readout/Passive function	Output/Active Function	Function Modifier
A	Analysis		Alarm		
B	Bumar, Combustion		User's choice	User's choice	User's choice
C	User's choice			Control	Close
D	User's choice	Differance, Differantial			Deviation
E	Votago		Sensor, primary element		
F	Flow, Flow Ratio	Ratio			
G	User's choice		Glass, Gauge, Viewing Device		
H	Hand				High
I	Current		Indicate		
J	Power	Scan			
K	Time, schedule	Time Ratio of change		Control station	
L	Leval		Light		Low
M	User's choice				Middle intermediate
N	User's choice		User's choice	User's choice	User's choice
O	User's choice		Orifice, Restriction		Open
P	Pressure		Point (Test connection)		
Q	Quantity	Intagrate, totalize	Integrate, totalize		
R	Radiation		Record		Run
S	Speed, Frequency	Safety		Switch	Stop
T	Temperature			Transmit	
U	Multivariable		Multifunction	Multifunction	
V	Vibration, Mochanical analysis			Valve, Damper, Louver	
W	Weight, force		Wal, Probe		
X	Unclassified	X-axis	Accasisary Devices, unclassified	Unclassified	Unclassified
Y	Event, State, Presence	Y-axis		Auxilary devices.	
Z	Position, dimension	Z-axis, safety instrumented system		Drivar, actuator, unclassifed final control element	

Figure 5.1 ISA tagging letter assignments.

method becomes limited. With today's complex SCADA software, the ISA tagging system cannot address all of the point names required in a typical application.

5.2 Purpose of Standardized Tagnames

A consistent naming convention should be developed for assigning descriptive tagnames to all of the points in the SCADA software. Points affected are the hardware input and output field signals, the internal virtual program points in the Programmable Process Controller (PPC) software and the database points used in the SCADA Operations Workstation (SOW) displays and software. Ideally, the tagname system should be applicable not only to the PPC programs but also to the Human–Machine Interface (HMI) software.

For example, a setpoint in the HMI software to which a user may assign an engineering value should be the same tagname that is used in the PPC control program. A setpoint tagname such as Tank_High_Level_Start could be used in the HMI application as well as the PPC program.

Older Programmable Logic Controllers (PLCs), such as Allen-Bradley PLC-5 and SLC-500 and the GE Fanuc 90–70 and 90–30 products, use a system of data types like integers and discretes. In these systems, tagnames would be added in the form of 'symbols' or 'descriptions' within the documentation of the program.

Today, controllers like the Allen-Bradley ControlLogix and the GE Fanuc PACSystem allow the user to use alphanumeric tagnames rather than data types with commented tagnames. The ControlLogix PAC allows the creation of a tagname such as 'Tank_High_Level_Setpoint' to represent the high-level setpoint for a control program. Both the HMI/SCADA workstation software and the ControlLogix program can use this same tagname, without any cross-referencing required.

5.3 Constructing Tagnames with Phrases

A common method of applying tagnames to points, both field and virtual, is to use a tagname made up of phrases or fields of characters. In the 'C' programming language, it is standard to concatenate fields to compose a variable name. In the same way, a tagname can be constructed using established phrases.

An initial approach might be to concatenate words to make up a tagname, taking advantage of the fact that the PPC programming software allows tagnames with up to 40 characters. Like the variable names in the 'C' programming language, related words can be 'strung together'. Consider some example tagnames using this method:

Line4ConveyorRunStatus
BoosterStationPump2DischargeValveOpenStatus
CollectorLine6JamFaultAlarm

While these tagnames are fully descriptive, they are very long and unmanageable. Using such tagnames to reference input and output field signals becomes impractical. Alternatively, abbreviations could be used instead of full words.

Using abbreviated phrases or fragments put together, a meaningful tagname can be created with far fewer characters; the abbreviated tagname still conveys the important information about the signal. Consider the following revised tagnames:

STNPump3DVOpen	Station Pump 3 Discharge Valve Open Status
HLPump2StartPB	High Lift Pump 2 Start Pushbutton
TankHILevel	Tank High Level Alarm

For readability, punctuation characters could be placed between phrases, such as the hyphen (-), underscore (_) or the period (.). Some software systems may accept the underscore, but not the hyphen; this will depend entirely on the specific software package being used. Modifying the above tagname examples would result in the following:

STN_Pump3_DV_Open	OR	STN_Pump3_DV_Open
HL_Pump2_Start_PB	OR	HL_Pump2_Start_PB
Tank_High_Level	OR	Tank_High_Level

Perhaps an improved method would be to define a tagname with fixed length fields or phrases in which each possible phrase for each field is defined. This is the method proposed in this book and a detailed explanation follows.

5.4 Tagname and Signal Naming Convention Structure

The Tagname Signal Naming Convention (TSNC) consists of 15 characters arranged into four fragments or elements; the elements are separated by an underscore, a hyphen or a period. Some SCADA HMI software may not allow the use of some of these delimiters, so the user will have to use whatever works best in their system.

The complete tagname follows a standard format, providing successive detail about the signal, its source and the specifics of the signals, in each fragment or element. Each field has defined alphanumeric codes or phrases which can be organized into 'dictionaries'; to construct a tagname, one assembles phrases from the various dictionaries to build a successively more detailed definition of the signal. This TSNC applies to both hardware or field signals and to the internal software or program points. Characters in the fourth field identify the nature of the signal and whether it is a field or a virtual type signal.

For example, a pump at the New Age Pumping Station would be identified with the initial fragments of NAP_PP05. The discharge valve for the pump would be tagged as NAP_PP05_DV. And the open status limits switch on the valve would complete the tagname as NAP_PP05_DV_SOD. The last fragment, SOD, is composed of a signal type, S, and a status field, OD.

5.4.1 Structure of the TSNC Tagname

The general format of the TSNC tagnames is as follows, using the underscore to separate the fragments of the tagname:

LLL_ EEEE_ CC_ TSS

where

LLL	is the Location within the overall system,
EEEE	is the major Equipment,
CC	is the Component of the equipment,
T	is the Type of Signal, including field or virtual,
SS	is the Signal Designation.

Note that the fourth fragment consists of two parts: a signal type (T) and a signal designation (SS); hence, this fragment serves to identify both the type of information and the signal detail.

5.4.2 Location Fragment

The **LLL** element identifies the geographic location of the equipment. These three alphanumeric characters define the physical location in the overall SCADA system. This fragment may be further divided into two characters, one for the location and the other character for the type of facility. For example, HLF could denote the High Lift Station of a water treatment plant; NA indicates New Age station and the P denotes the pumping station; NAR indicates New Age and the R denotes the reservoir.

5.4.3 Equipment Fragment

The **EEEE** element identifies the actual equipment or device, including the unit number. This fragment can be subdivided with some characters for the equipment and some characters for the number. Two of the four characters can be used to designate the unit number of equipment; three letters for the equipment results in one letter for the unit and four letters could be used to designate the complete facility. Consider the following examples:

HLPx	High Lift Station Pump x
BPVx	Bypass Valve x
BLDG	Building (no number)
FLxx	Filter xx, range 1–99
PPxx	Pump xx, range 1–99

5.4.4 Component Fragment

The **CC** element identifies the component of subsystem of the main equipment or device. If the entire equipment is being referenced as a single unit, then the CC field could be ‘00’. Typically, the CC field indicates the part of the equipment being referenced such as DV for discharge valve, IV for inlet valve and DG for drain gate. Some examples of component fragments are shown below:

RWP2_DV	Raw Water Pump 2, Discharge Valve
RWL1_FT	Raw Water Line 1, Flow Transmitter
TWP4_PT	Treated Water Pump 4, Pressure Transmitter

One additional use for the component element is to designate one of several unit signals associated with the basic equipment element. For example, a motor may have four Resistive Thermal Device (RTD) temperature sensors associated with it, so the component field can be the numbered sensor:

HLF_HLP5_01_ATA	pump #5 temperature sensor #1
HLF_HLP5_02_ATA	pump #5 temperature sensor #2

HLF_HLP5_03_ATA	pump #5 temperature sensor #3
HLF_HLP5_04_ATA	pump #5 temperature sensor #4

5.4.5 *Signal Type and Designation*

The **TSS** element combines two parts: the signal type, T, and the signal designation, SS. Since one of the goals of the TSNC is to adopt a universal naming system for both the field hardware signals and the internal program points, the signal type, T, indicates which of several point types apply to the tagname.

The **T** element, or Signal Type, will be one of the following:

S	discrete status input	X	virtual discrete/boolean input point
A	discrete alarm status input	Y	virtual discrete/boolean output point
D	discrete on/off output	Z	virtual boolean flag point
Q	analog value input	N	virtual analog input; internal value
K	analog value output	R	virtual analog output; internal value

Note that the virtual points are those for which there is no hardware or field counterpart; it is not wired to anything in the field. For example, if the two analog input signals, HLF_RWL1_FT_QFI and HLF_RWL2_FT_QFI, are added together, the sum may be stored in a virtual total, HLF_RWL0_FI_NFI.

The **SS** element, or Signal Designation, identifies the specific signal associated with the specified equipment. This fragment relies heavily on the ISA naming convention. While this ISA system does identify the nature of the signal, it does not identify the location or the type of equipment being referred to. The ISA tagging system was illustrated in Section 5.1 previously.

Some examples of signal designation fragments are shown below:

RWP3_SS_SRC	Remote Control Selector
MIV2_EV_DON	Open Control on valve
AST1_LS_ALH	Alum Tank High Level Sensor
BSP2_PF_ZRM	Pump 2 Remote Manual Selected

5.5 Tagname Fragment Dictionaries

As stated earlier, the TSNC uses a four fragment system in which each fragment identifies one part of the complete tagname. A series of 'dictionaries' may prove useful in constructing tagnames for a SCADA system. As the system is designed and developed, additional fragments may be added to these dictionaries. At the end of the project, there will be a complete set of listings for all fragments of the tagnames used in the SCADA project.

Each of the four fragments or elements of the tagname will have a list of assigned phrases or definitions. Appendix 'A' contains a series of dictionaries for the tagname fragments, based upon the water treatment plant application. Of course, this system may be modified and tailored in any way desired; this is meant as a design methodology and provide a starting point for tagname construction.

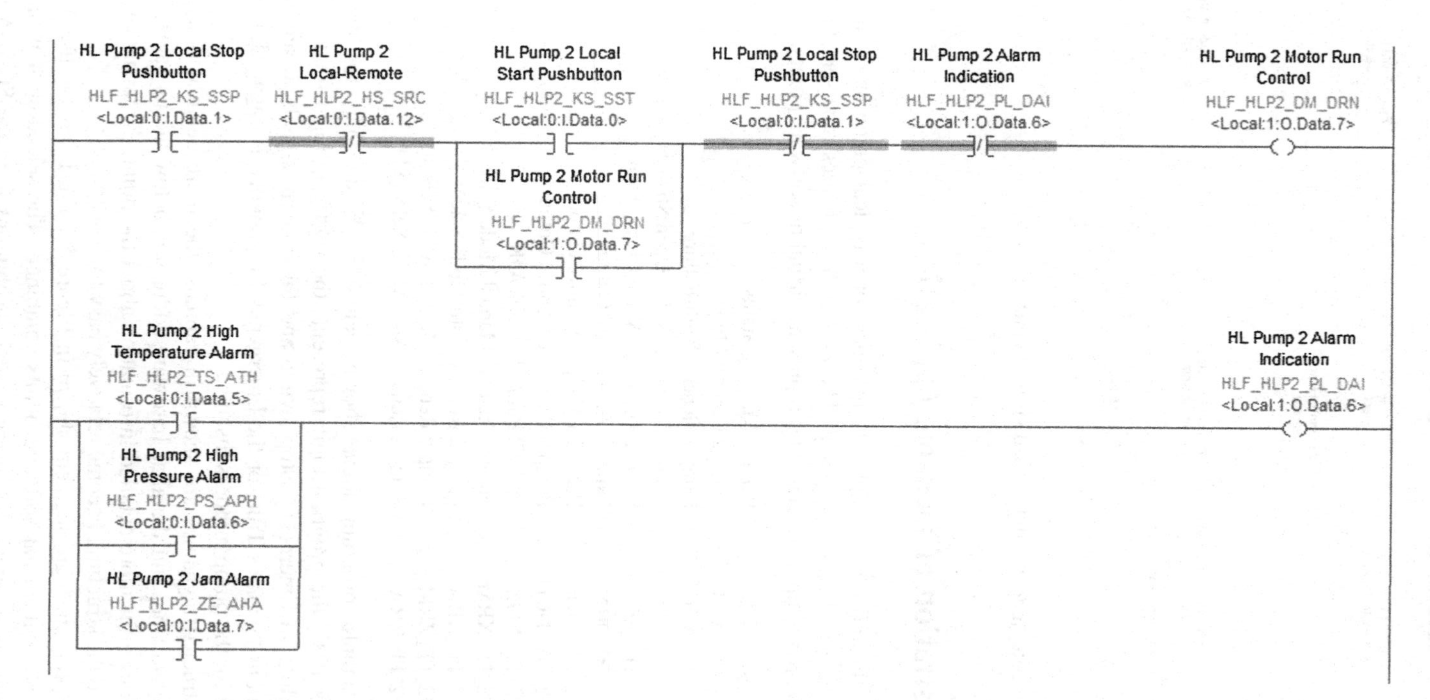

Figure 5.2 Program sample with operand descriptions.

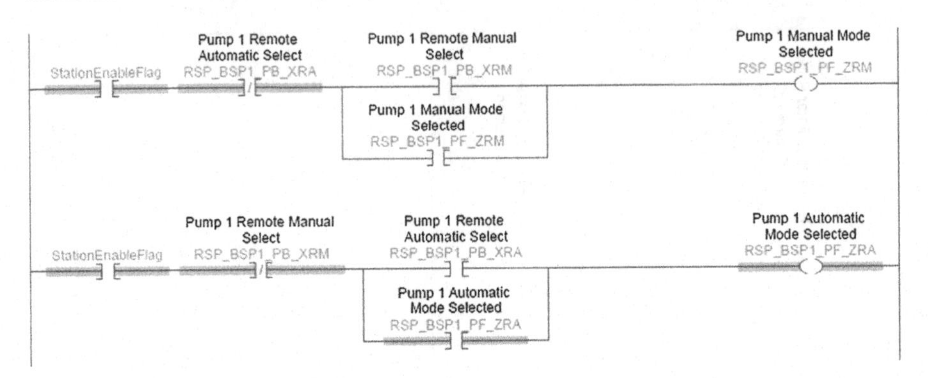

Figure 5.3 Program sample with field and virtual points.

5.6 Illustration of Tagname Application

This section illustrates an example of tagnames for a typical pump and some program logic associated with that pump. The first eight tagnames represent field I/O points, and the remaining five tagnames represent virtual or software points.

LLF_RWP2_00_SRN	Pump 2 Running Status
LLF_RWP2_00_SCM	Pump 2 Control Mode
LLF_RWP2_00_AGF	Pump 2 General Fault/Failure
LLF_RWP2_DV_SOD	Pump 2 Discharge Valve Open Status
LLF_RWP2_DV_SCD	Pump 2 Discharge Valve Close Status
LLF_RWP2_00_DRN	Pump 2 Run Control Output
LLF_RWP2_DV_DON	Pump 2 DV Open Control Output
LLF_RWP2_DV_DCD	Pump 2 DV Close Control Output
LLF_RWP2_PF_YOK	Pump 2 Pump Available/OK
LLF_RWP2_PF_XRM	Pump 2 Remote Manual Mode Select
LLF_RWP2_PF_XRA	Pump 2 Remote Auto Mode Select
LLF_RWP2_PF_ZRM	Pump 2 Remote Manual Mode Selected
LLF_RWP2_PF_ZRA	Pump 2 Remote Auto Mode Selected

In the example program logic shown below in Figure 5.2, for the A-B ControlLogix PLC, the program logic represents the start/stop logic for a high lift pump, including an Emergency Stop switch and three alarm signals. If any of the alarm conditions becomes True, or the Emergency Stop switch is engaged, then the control logic for the pump motor is disabled.

The constructed tagnames are shown in capitals, the more detailed description is above the tagname, and the alias hardware address is below the tagname. While the descriptions offer more detailed information about the point, the tagnames, once accustomed to the structure, become relatively easy to understand.

The next program logic segment shown in Figure 5.3 illustrates the use of both field I/O points and virtual software points combined. The selection of manual or automatic mode is accomplished using virtual pushbuttons from the SCADA workstation, and the selected mode is stored in virtual 'selected' mode points.

6 Developing the Application Program Databases

Objectives

- Define the basic field I/O data types
- Describe the data types found in controller programs
- Explain how the database spreadsheets are created
- Present a practical method for creating database spreadsheets

The program database is not a single file, but rather a collection of database files, one for each Programmable Process Controller (PPC) and SCADA Operations User Workstation (SOW) in the SCADA system. In the previous chapter, a tagging system was described which can be applied to both field and virtual points in a SCADA system. Each PPC would require its own hardware and software points lists or databases; likewise, the SCADA application executing on the SOW requires a software points list or database of all data to and from the PPCs. Following are sections which address each of the database categories.

6.1 Review of Data Types and Databases

SCADA systems process data in different formats or types, and the information is maintained in one or more databases. In the case of the PPC, the database is kept in RAM or the working memory of the PPC. For the SOW, the databases are kept both in RAM and on the hard drive of the computer. It is worth noting that the IEC 61131-3 programming standard has established these data types for PPC programming. Before considering the methodology of creating and maintaining databases for SCADA systems, a brief review is offered on the data types used in SCADA systems.

6.1.1 Raw Field Signal Data

The PPC processes two basic field signal types, each of which can be an input or an output relative to the PPC. At the PPC I/O module level, there are four basic signal types: discrete input and output, analog input and output. Figure 6.1 illustrates the formats of these two basic data types. An overview of the signal processing performed by the PPC is provided in Section 1.2.

Designing SCADA Application Software. DOI: http://dx.doi.org/10.1016/B978-0-12-417000-1.00006-3

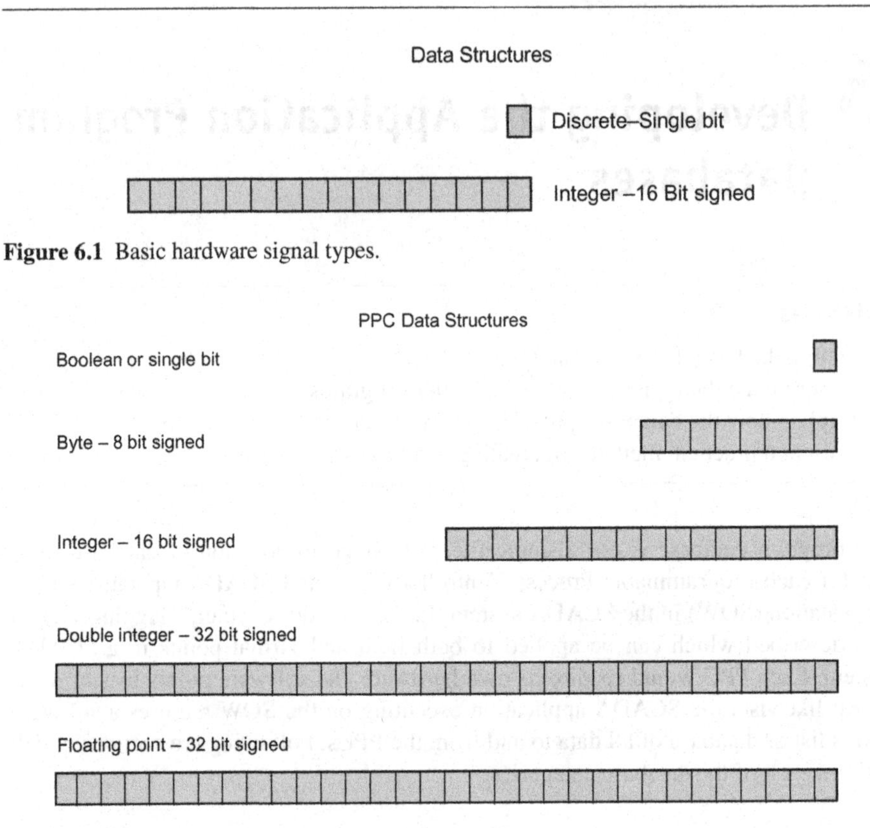

Figure 6.1 Basic hardware signal types.

Figure 6.2 Typical PPC data types available.

The discrete input or output signal is a two-state signal which is either on or off, true or false, one or zero; this data is represented by a single Binary Digit, or 'Bit'.

The analog input or output signal has a range of values between two limits; for example, a speed signal from 0 to 5000 rpm, a level signal from 0 to 3.4 m, or a chlorine residual from 0 to 5.0 ppm. A group of 16 bits form the 'Integer' or 'Word' data type, which can accommodate any value between 0 and 65,535 unsigned, or −32,768 to +32,767 signed. It is then up to the PPC programmer to convert this analog data into useful engineering values, which is the topic of a later section.

6.1.2 PPC Data Types

In the past, the integer format was 16 bits signed, as shown previously; today, most PPCs support both the original Integer and the Double Integer, which is 32 bits in length. Thus, analog inputs and outputs must be specified as to which of the two analog formats is to be used.

Internal to the PPC, the database points may be in a number of formats, depending upon the application of the signal or program point. Figure 6.2 provides an illustration showing the most common data types that would be found in today's PPC.

6.1.3 SOW Program Data Types

The SCADA Workstation, with its process graphic displays and other user interface data, typically uses all of the basic data types shown in the previous section. One key difference is that the SOW database includes the field I/O points from all of the PPCs, since information about all of the process areas must be made available to the user.

The SOW supports the field I/O data points like those shown previously; they also support internal or 'Memory' type points which reside only in the Human–Machine Interface (HMI) application software. This is similar to the idea that the PPC supports internal software points, as well as hardware or field I/O points.

The basic database for the SCADA Workstation application typically contains the same points as found in the Controllers; however, the SOW includes the points from all of the PPCs, since the workstation must display information from all of these controllers. The communication drivers or servers collect data values from each of the controllers, and maintain a current record of each point in the system in a process database. As the controllers perform their programmed operations, data is being passed on to the workstation's database. As will be explained later in this chapter, the SOW database must include all of the field I/O points from all of the PPCs, plus all of the virtual points in the controllers.

A workstation will also have internal or 'memory' type data points in a number of formats. Just as there are PPC integer and floating point data points, which reference the PPC database points, there are also internal or virtual points for the SOW database. SCADA software in the workstation may use any or all of the following additional data types:

- Memory Discrete
- Memory Integer (16-bit and 32-bit)
- Memory Real or Floating Point
- Message – text based data
- History Trend

These additional data types are used for internal database points and will vary in number and format from one software package to another. The point being made here is that in addition to the various data types used in the PPC, there are additional data types used in the SOW database. This will be covered in more detail later in this chapter.

6.2 Using Spreadsheets to Create Points Lists

Creating points lists for both the hardware I/O signals and the internal program points can best be done using the spreadsheet, such as Microsoft Excel. The spreadsheet allows the creator to organize the information into any number of columns, add text headings, and formatting to make the spreadsheets presentable. These spreadsheets also serve as 'Shop Drawings' for contracts as the points lists clearly show how each PPC is organized and configured.

The layout of the spreadsheet should include, as a minimum, the following fields:

Tagname	TSNC defined name for the signal
Point Description	text description of the signal
Point Type	DI, DO, AI, AO
Hardware Address	hardware reference for program
Module Code	catalog number for the I/O module

6.2.1 Basic Spreadsheet Template

Figure 6.3 shows a sample spreadsheet layout using the above fields, with sample points entered into the columns. For each entry, the standardized tagname is shown together with its full description, followed by the details of the point type, point address, and module code.

Additional information about the PPC can be included in the heading area, such as:

- Name of controller – application name
- IP address on network

6.2.2 Populating Points List Spreadsheets

Using the template shown previously, the fields can be populated by entering the details of the controller points, such as the status inputs and control outputs. Information about the Terminal Numbers will not be known until commissioning, when the electrician completes the wiring to the PPC modules. If changes are made during installation and/or commissioning, then the spreadsheets can be readily changed to document the changes.

Typically, many I/O modules can be included on one spreadsheet. For modules with 16 inputs or outputs, only two modules will fit; for modules with four or eight signals, then more modules can be accommodated. The spreadsheet should be formatted so that the sheets can be printed on standard sized paper. The number of modules on a given sheet is based upon what can fit within the space of the formatted spreadsheet.

One of the benefits of using the spreadsheet program is that the auto incrementing of addresses can make address entry faster and simpler. For example, if the first two or three addresses are entered, then the remaining addresses for the module can be created using the select and drag feature. Also, tagnames and descriptions can be copied and pasted and then edited, rather than typing each entry individually.

Having prepared a points list manually, the following example in Figure 6.4 illustrates how the information might be tabulated into columns. This information would then be entered into the spreadsheet; the addresses and module codes are based upon the SLC-500 PLC.

Figure 6.5 provides an illustration of a spreadsheet that has been populated with some of the I/O discrete signals for a specific PPC. The two modules shown include space for all eight inputs and outputs. As can be seen, those points which are assigned have tagnames and descriptions entered, while the blank fields represent spare entries.

Name of SCADA Project **Name of PPC Process**		*Controller Name:* *Network IP:*			
Tagname	Point Description	Point Type	Hardware Address	PLC Module Prod. Code	Terminal Numbers

Figure 6.3 Blank spreadsheet template.

Tagname	Point Description	Point Type	H/W Address	Module Code
LLF_RWP1_00_SRN	LL Pump 1 Running Status	DI	I:7/0	1746-IA8
LLF_RWP1_00_SCM	LL Pump 1 Remote Mode	DI	I:7/1	1746-IA8
LLF_RWP1_00_AGF	LL Pump 1 General Fault	DI	I:7/2	1746-IA8
LLF_RWP1_PS_APL	LL Pump 1 Low Pressure Alarm	DI	I:7/3	1746-IA8
LLF_RWP1_PS_APH	LL Pump 1 High Pressure Alarm	DI	I:7/4	1746-IA8
LLF_RWP1_DV_SCM	LL Pump 1 DV Remote Mode	DI	I:7/5	1746-IA8
LLF_RWP1_DV_SCD	LL Pump 1 DV Close Status	DI	I:7/6	1746-IA8
LLF_RWP1_DV_SOD	LL Pump 1 DV Open Status	DI	I:7/7	1746-IA8

Figure 6.4 Sample discrete input module configuration.

Note that the controller code name and IP address have been added to the header of the spreadsheet. The filename, data and page number could also be included here. Additional spreadsheets would be added to show the configuration and contents of all of the modules in the controller in Figure 6.5.

6.2.3 *Creating Spreadsheet Workbook Files*

As has been illustrated, single spreadsheets containing the details of the various I/O module can be created to document all of the signals in a given PPC or controller; with some additional sheets, a cover sheet and summary information can be added to create a complete spreadsheet (i.e. Excel) workbook file.

Figure 6.6 is a sample cover spreadsheet which contains summary information about the controller, including the number of signals of each type currently assigned, and the number of spare points in the modules. This summary data can be useful for future expansion of the project.

Since this is a spreadsheet, the ‘Used’ and ‘Total’ columns can be entered, and the ‘Spare’ column can be calculated. An additional percentage column could be included to the right of the ‘Total’ column to give an indication of the percentage of available points in each module. If additional signals are to be added to this controller, the programmer can see how many spare points of each type are available in the current configuration.

Another sheet that might be useful is a configuration sheet, as illustrated in Figure 6.7; this would show the current module locations and types of each module in each of the racks or chassis. This sheet would also indicate any spare slots in the controller for adding more modules if required.

6.3 Developing the Controller Database

The database points within a typical PPC contain both real, or field I/O points, and virtual, or software points. The software points, in turn, consist of virtual program points used in the logic and ‘support points’ used for executing the logic. Every point begins with an assigned tagname based upon a tagname structure such as described in Chapter 5.

Water Treatment Plant **Raw Water Pumping Station**		*Controller Name:*		RWPSTN	
		Network IP:		192.168.205.141	
Tagname	Point Description	Point Type	Hardware Address	PLC Module Prod. Code	Terminal Numbers
LLF_RWP1_00_SRN	*LL Pump 1 Running Status*	DI	I:7/0	1746-IA8	
LLF_RWP1_00_SCM	*LL Pump 1 Remote Mode*	DI	I:7/1	1746-IA8	
LLF_RWP1_00_AGF	*LL Pump 1 General Fault*	DI	I:7/2	1746-IA8	
LLF_RWP1_PS_APL	*LL Pump 1 Low Pressure Alarm*	DI	I:7/3	1746-IA8	
		DI	I:7/4	1746-IA8	
		DI	I:7/5	1746-IA8	
		DI	I:7/6	1746-IA8	
		DI	I:7/7	1746-IA8	
LLF_RWP1_DM_DRN	*LL Pump 1 Motor Run Control*	DO	O:4/0	1746-OA8	
LLF_RWP1_DV_DON	*LL Pump 1 Disch. Valve Open*	DO	O:4/1	1746-OA8	
LLF_RWP1_DV_DCE	*LL Pump 1 Disch. Valve Close*	DO	O:4/2	1746-OA8	
		DO	O:4/3	1746-OA8	
		DO	O:4/4	1746-OA8	
		DO	O:4/5	1746-OA8	
		DO	O:4/6	1746-OA8	
		DO	O:4/7	1746-OA8	

Figure 6.5 Illustration of populated spreadsheet.

Water Treatment Plant Raw Water Pumping Station	*Prepared By:*	*Date:* *Filename:*

Summary of I/O Points:

		Used	**Spare**	**Total**
Discrete Inputs	:	56	16	72
Discrete Outputs	:	37	11	48
Analog Inputs	:	14	10	24
Analog Outputs	:	5	3	8

Figure 6.6 Suggested cover page for spreadsheet workbooks.

Water Treatment Plant Raw Water Pumping Station	*Prepared By:*	*Date:* *Filename:*

	Slot	*Module*	*Description*
Rack 0 :	1	1747-L552	*CPU Processor*
	2	1746-IA16	*Digital Input Card - 16 Point*
	3	1746-IA16	*Digital Input Card - 16 Point*
	4	1746-IA16	*Digital Input Card - 16 Point*
	5	1746-IA16	*Digital Input Card - 16 Point*
	6	1746-IA16	*Digital Input Card - 16 Point*
	7	1746-OA16	*Digital Output Card - 16 Point*
	8	1746-OA16	*Digital Output Card - 16 Point*
	9	1746-OA16	*Digital Output Card - 16 Point*
	10	1746-OA16	*Digital Output Card - 16 Point*
Rack 1 :	1	1746-NI4	*Analog Input Card - 4 Point*
	2	1746-NI4	*Analog Input Card - 4 Point*
	3	1746-NI4	*Analog Input Card - 4 Point*
	4	1746-NI4	*Analog Input Card - 4 Point*
	5	1746-NI4	*Analog Input Card - 4 Point*
	6	1746-NI4	*Analog Input Card - 4 Point*
	7		
	8		
	9	1746-NO4I	*Analog Output Card - 4 Point*
	10	1746-NO4I	*Analog Output Card - 4 Point*

Figure 6.7 Suggested configuration page for spreadsheet workbooks.

6.3.1 Hardware Field I/O Signals

Each of the PPCs will be interfaced to a number of field signals; these signals must be identified and categorized as to type and purpose. In summary, the types are:

DI	Discrete Input	on or off status input
DO	Discrete Output	on/off control output
AI	Analog Input	continuous range of values input
AO	Analog Output	variable control value to device

PPCs handle inputs and outputs in groups via interface cards or modules. A typical discrete input module might handle 8, 16 or 32 discrete inputs; likewise, a discrete output module might handle 8, 16 or 32 discrete outputs. Analog modules handle a smaller number of signals, such as 4 or 8 input or output signals. Hence, a PPC will include any number of input and output modules.

Signals related to a single piece of equipment are usually grouped together on the same module; for example, all of the discrete inputs for a pump would be grouped together on consecutive inputs of a single module. This approach to organizing the points makes it simpler to locate I/O points within the points list and to locate all signals related to a piece of equipment.

Each such field I/O point would have attributes including: tagname, description, hardware address, point type, module catalog number, and perhaps cross reference to an ISA tagname and terminal numbers or wire numbers. In Figure 6.8, a sample spreadsheet is shown for discrete inputs to a PPC. The spreadsheet is organized into columns to show the attributes mentioned earlier. Two additional columns that can be quite useful are the Terminal Numbers and the Alternate Tag Reference.

The *Terminal Numbers* column provides a connection reference for wiring of the input signals to the module. For both initial installation and later troubleshooting, the terminal strip numbers make it easier to locate the connection point for each of the signals.

The *Tag Reference* column might be included if another tagging system is being used. Some projects include an ISA tagname, as shown in the spreadsheet; including this reference in the points list offers a quick cross reference, as the ISA tagname is often found on the contract drawings.

The second illustration in Figure 6.9 is one for the analog input and output modules. The spreadsheet includes the analog inputs for the low lift pump discharge valves, and the analog outputs for controlling the valves. Like the previous examples, all of the same columns are included, identifying the tagname, hardware address and location, module type, and so on.

The input module type is a 1746-NI8, 8-channel analog input type. The analog inputs are referenced as points 0 through 7. The output module type is a 1746-NO4I which is a 4-channel analog output with points 0 through 3. These examples are based upon the A-B SLC-500 PLC.

Analog signals must be converted to numeric values in binary, typically in Integer or Real format. Further processing of analog signals may be required depending upon how the signals are used in the PPC control program.

Water Treatment Plant **Raw Water Pumping Station**		*Controller Name:* RWPSTN *Network IP:* 192.168.205.141				
Tagname	Point Description	Point Type	Hardware Address	PLC Module Prod. Code	Tag Reference	Terminal Numbers
LLF_RWP2-00-SRN	*LL Pump 2 Running Status*	DI	I:7/0	1746-IA8	GB-020	1120-07
LLF_RWP2-00-SCM	*LL Pump 2 Remote Mode*	DI	I:7/1	1746-IA8	HI-020	1120-10
LLF_RWP2-00-AGF	*LL Pump 2 General Fault*	DI	I:7/2	1746-IA8	XS-020	1120-11
LLF_RWP2-PS-APL	*LL Pump 2 Low Pressure Alarm*	DI	I:7/3	1746-IA8	PSL-021	1120-12
LLF_RWP2-PS-APH	*LL Pump 2 High Pressure Alarm*	DI	I:7/4	1746-IA8	PSH-022	1120-13
LLF_RWP2-DV-SCM	*LL Pump 2 DV Remote Mode*	DI	I:7/5	1746-IA8	HI-024	1120-14
LLF_RWP2-DV-SCD	*LL Pump 2 DV Close Status*	DI	I:7/6	1746-IA8	ZSC-024	1120-15
LLF_RWP2-DV-SOD	*LL Pump 2 DV Open Status*	DI	I:7/7	1746-IA8	ZSO-024	1120-16
LLF_RWP3-00-SRN	*LL Pump 3 Running Status*	DI	O:4/0	1746-OA8	GB-060	1130-00
LLF_RWP3-00-SCM	*LL Pump 3 Remote Mode*	DI	O:4/1	1746-OA8	HI-060	1130-01
LLF_RWP3-00-AGF	*LL Pump 3 General Fault*	DI	O:4/2	1746-OA8	XS-060	1130-02
LLF_RWP3-PS-APL	*LL Pump 3 Low Pressure Alarm*	DI	O:4/3	1746-OA8	PSL-061	1130-03
LLF_RWP3-PS-APH	*LL Pump 3 High Pressure Alarm*	DI	O:4/4	1746-OA8	PSH-062	1130-04
LLF_RWP3-DV-SCM	*LL Pump 3 DV Remote Mode*	DI	O:4/5	1746-OA8	HI-064	1130-05
LLF_RWP3-DV-SCD	*LL Pump 3 DV Close Status*	DI	O:4/6	1746-OA8	ZSC-064	1130-06
LLF_RWP3-DV-SOD	*LL Pump 3 DV Open Status*	DI	O:4/7	1746-OA8	ZSO-064	1130-07

Figure 6.8 Sample discrete I/O points list.

Water Treatment Plant
Raw Water Pumping Station

Controller Name: RWPSTN
Network IP: 192.168.205.141

Tagname	Point Description	Point Type	Hardware Address	PLC Module Prod. Code	Tag Reference	Terminal Numbers
LLF_RWP1-IE-QII	*LL Pump 1 Current*	AI	I:1.0	1746-NI8	II-010	1160-00
LLF_RWP2-IE-QII	*LL Pump 2 Current*	AI	I:1.1	1746-NI8	II-020	1160-01
LLF_PWL2-LT-QLI	*Low Lift Well 2 Level*	AI	I:1.2	1746-NI8	LIT-050	1160-02
LLF_RWP3-IE-QII	*LL Pump 3 Current*	AI	I:1.3	1746-NI8	II-060	1160-03
		AI	I:1.4	1746-NI8		1160-04
LLF_RWP1-DV-QZI	*LLP1 Discharge Valve Position*	AI	I:1.5	1746-NI8		1160-05
LLF_RWP2-DV-QZI	*LLP2 Discharge Valve Position*	AI	I:1.6	1746-NI8		1160-06
LLF_RWP3-DV-QZI	*LLP3 Discharge Valve Position*	AI	I:1.7	1746-NI8		1160-07
LLF_RWP1-DV-KSP	*LLP1 Discharge Valve Setpoint*	AO	O:2.0	1746-NO4I		1150-00
LLF_RWP2-DV-KSP	*LLP2 Discharge Valve Setpoint*	AO	O:2.1	1746-NO4I		1150-01
LLF_RWP3-DV-KSP	*LLP3 Discharge Valve Setpoint*	AO	O:2.2	1746-NO4I		1150-02
		AO	O:2.3	1746-NO4I		1150-03

Figure 6.9 Sample analog I/O points list.

6.3.2 Software Program Points

The Hardware Field I/O Signals represent physical signals wired to input and output modules in the PPC. Internal to the PPC application program are virtual or software points used in the program logic. These points require tagnames and attributes the same as the field I/O points, but there is no physical correspondence to these points. Consider the following explanation.

All database points in the PPCs require an alphanumeric tagname to identify the software point; various tagname characteristics are then configured for that tagname. Points may be field (meaning hardware) or virtual (meaning internal software), depending upon the nature of the point. Both the A-B ControlLogix and the GE Fanuc PACSystem use tag-based points in the PPC programs.

A PPC database point, therefore, consists of an identifying tagname with additional identifying characteristics such as the following:

Tagname	Alphanumeric string to identify the point
Description	Optional alphanumeric description
Type of Point	Virtual (Base) or Field (Alias)
Alias or H/W	For field points, the hardware reference
Data Type	Boolean, Integer, Real, etc.

The distinction between Virtual and Field above differentiates between internal software and physical hardware point types. Note that all points in the PPC database include the characteristics listed previously. Hence, all database points defined in the PPC application program would consist of points with an identifying tagname and the typical characteristics indicated. The specific characteristics will vary from one PPC manufacturer to another, but all points have a tagname and point type.

To illustrate the concept and use of virtual points, consider the example code segments in Figures 6.10 and 6.11. In the first example, there are three field or alias discrete input signals which are being combined such that if all conditions are True, then the output internal virtual point is True. This logic sets a virtual boolean flag to indicate that there are no alarm conditions and that the pump is in the Remote mode.

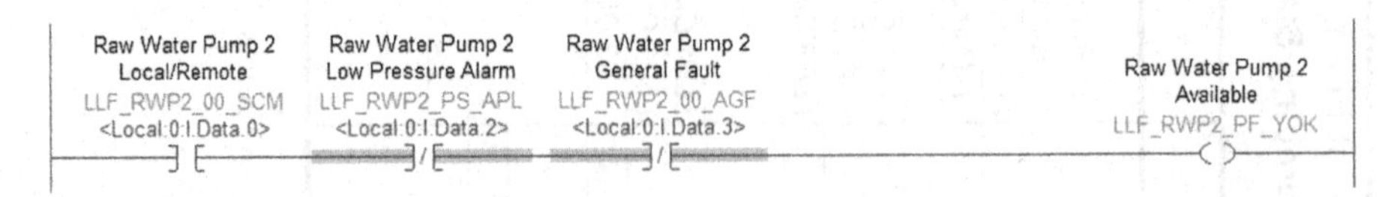

Figure 6.10 ControlLogix program logic example.

Motor remote Selected
LLF_RWP2_KS_SRM
I:1/6
Motor Temperature Alarm
LLF_RWP2_TE_ATA
I:1/4
Motor Bearing Alarm
LLP_RWP2_DM_APF
I:1/5
Motor Available
LLF_RWP2_PF_YRD
B3/16

Figure 6.11 SLC-500 program logic example.

In the second example, there are three field points being logically-ANDd so as to combine the necessary conditions to operate the pump. A virtual discrete point, LLF_RWP2_PF_YRD, is set or cleared to indicate that the pump is available for use. This latter virtual signal could then be used to enable the remaining logic for the pump; if any of the required conditions becomes False, then the pump logic becomes disabled.

6.3.3 Application of Program Points in Logic

Program points can be incorporated into program logic to produce other program points. Another example of the use of program virtual boolean flags is the setting of the remote mode of control. Most SCADA systems allow for both Manual and Automatic modes of control. Virtual boolean flags are used to indicate which of these two modes is currently selected. In the following program examples, both the ControlLogix and SLC-500 versions of logic are shown. Refer to Figures 6.12 and 6.13 for the logic illustrations.

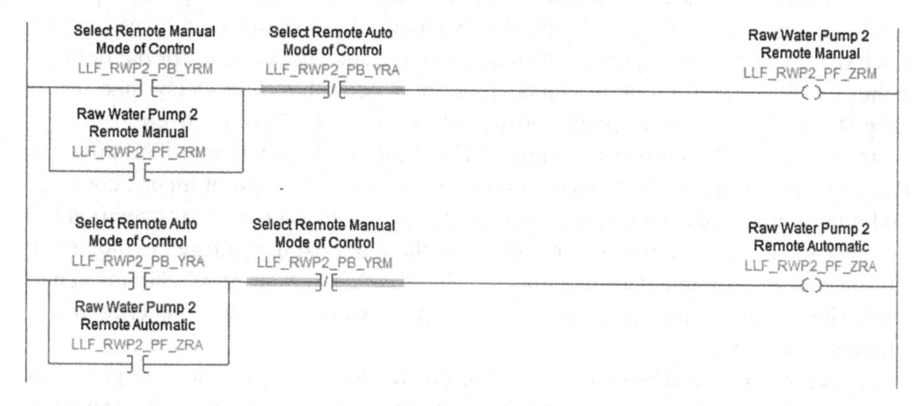

Figure 6.12 Sample RSLogix 5000 software point logic.

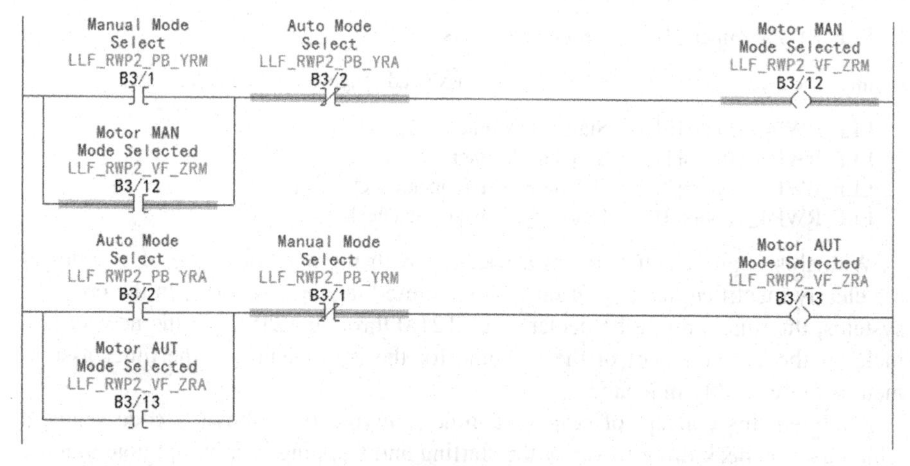

Figure 6.13 Sample RSLogix 500 software point logic.

The selection pushbuttons for Manual and Automatic modes of control are derived from virtual points in the SCADA Operations Workstation software: Manual and Automatic mode select LLF_RWP2_PB_YRM and LLF_RWP2_PB_YRA. When one of these two points becomes True, the logic causes the corresponding virtual mode indicator flag to become True (e.g. LLF_RWP2_PF_ZRM or LLF_RWP2_PF_ZRA). Note that the interlocking of the virtual pushbuttons ensures that both modes cannot be selected simultaneously. This is good programming practice, and in fact, would be implemented in electrical circuits if the controls were hardwired.

6.3.4 Internal Program Support Points

Finally, there are often virtual program points used to perform the program logic tests. Timers are used extensively to check that motors start and stop within an allowed time period, based upon the command to start or stop and the change in the field input status. Counters may be used to count objects passing by a sensor on a conveyor line. Each of these internal support points requires a tagname, the same as for the field I/O points and the virtual program points. There may be both a Start Delay timer and a Stop Delay timer, identified previously as 'T4:41' and 'T4:42', respectively.

In addition to the normal assembly of field and virtual points in a PPC program, there are also what may be termed 'support points' which consist of timers, counters, and other point types. Timers are often used to verify that a particular operation has completed, such as a pump motor starting. If the timer expires before the motor status matches the commanded state, then a 'Motor Failed to Start/Stop' alarm is generated. This is an example of a virtual support point which is used in the checking of process operations.

Arrays of timers and/or counters may be created for this type of support points for programs. For example, referring to the start and stop check timers above, an array of timers such as:

LLF_RWP4_Timers[4]   Group of 4 timers

might be created from which the following assigned usage could be made:

LLF_RWP4_Timers[0]   Start check timer
LLF_RWP4_Timers[1]   Stop check timer
LLF_RWP4_Timers[2]   Discharge valve open check timer
LLF_RWP4_Timers[3]   Discharge valve close check timer

With this approach, the timers associated with the specific piece of equipment are clearly identified, as they fit into the structured tagname scheme. In the previous systems, the timers might be declared as: T21:0 through T21:3; but the new system includes the essential part of the tagname for the equipment and the timer assignment is more readily indicated.

Applying this concept of structured timers, Figure 6.14 shows a code segment which uses a check timer to verify the starting and stopping of a pump; note that the timers used in the logic are part of an array of timers.

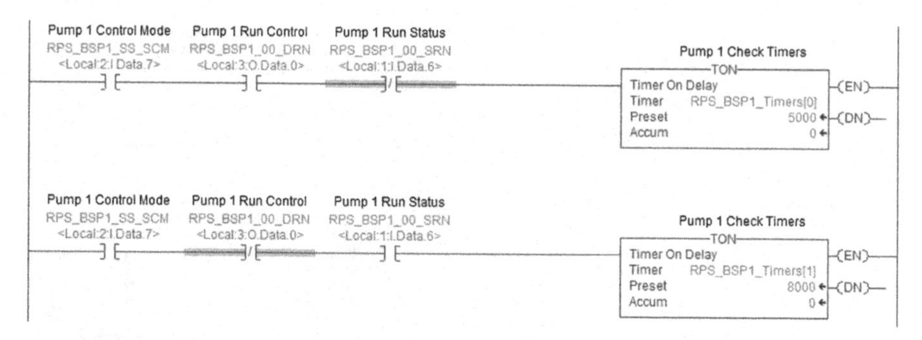

Figure 6.14 Use of structured array timers in program logic.

Equipment runtime, a common application for timers and counters, might likewise be created as arrays:

LLF_RWP4_Runtime[3]

in which the assignments could be:

LLF_RWP4_Runtime[0]	Runtime seconds
LLF_RWP4_Runtime[1]	Runtime minutes
LLF_RWP4_Runtime[2]	Runtime hours

Again, this approach incorporates the basic tagname for the equipment with the runtime counter tagname. Each of the 'runtime' elements would be declared as a retentive timer or counter type to accumulate the runtime for the equipment.

The use of structured array counters can be applied to the runtime counters, as shown in Figure 6.15 code segment for accumulating runtime in Hours and Tenths of Hours.

The assignment of these support points need to be documented along with the hardware and program points for a program. Using a structured tagname system allows all points to have a consistent naming convention which makes identifying the points considerably simpler.

A spreadsheet for software or virtual program points does not require columns for addresses and terminal numbers; instead, columns are required for data point type and internal versus external reference. Creating a spreadsheet for these virtual program points, including the support type points, might appear as shown in Figure 6.16.

In this illustration, there is no reference to the physical world (i.e. input and output signals) since these are internal program points. These points do, however, still require a tagname, description and definition of type of point. Note that the point type may be other than those listed earlier in Figure 6.2 describing the typical PPC data types.

Note in the example that the check timers and runtime counters are structured as arrays for more efficient programming; these points could be configured as individual points, or as arrays as shown.

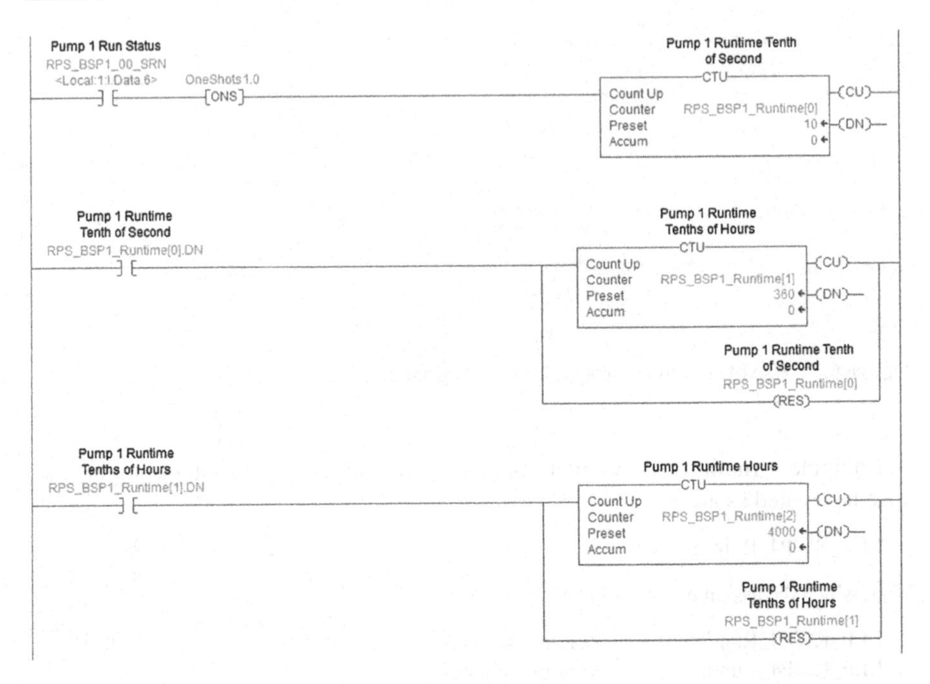

Figure 6.15 Use of structured counters in program logic.

6.3.5 *Procedures for Creating Controller Spreadsheet Database*

In Section 6.2, the use of spreadsheets was proposed as a way of creating points lists: tabulated listings of all field and virtual or program points used in a PPC program database. The purpose of these tabulated listings is twofold:

1. Organize all of the database points into listing by category, such as field I/O, internal program and virtual support points.
2. Produce a detailed document of the database for each controller in the SCADA project.

PPC or controller programming is typically done by creating points as they are needed in the development of the program logic; tagnames are created 'on the fly' with no real naming convention. If the original programmer is the only person that would ever have to work with this software, then that may be acceptable. In the real world, however, the SCADA project outlives the original programmer, and other people are involved in the addition and expansion of these software projects. For anyone other than the original programmer, the 'organization' of the database and the tagname usage may appear disorganized and perhaps even difficult to understand.

If, on the other hand, the database is developed and structured around a tag naming convention and the points are organized in a logical manner, then the final program database can be understood by any competent programmer, especially if the

Water Treatment Plant Raw Water Pumping Station			
Tagname	Description	Base / Alias	Point Type
LLF_RWP1_PF_YOK	*LL Pump 1 Available*	Base	Boolean
LLF_RWP1_PB_YRM	*LL Pump 1 Select Remote Manual*	Base	Boolean
LLF_RWP1_PB_YRA	*LL Pump 1 Select Remote Auto*	Base	Boolean
LLF_RWP1_PF_ZRM	*LL Pump 1 Remote Manual Mode*	Base	Boolean
LLF_RWP1_PF_ZRA	*LL Pump 1 Remote Auto Mode*	Base	Boolean
LLF_RWP1_Timers[0]	*LL Pump 1 Start Check Timer*	Base	Timer
LLF_RWP1_Timers[1]	*LL Pump 1 Stop Check Timer*	Base	Timer
LLF_RWP1_Timers[2]	*LL Pump 1 DV Open Check Timer*	Base	Timer
LLF_RWP1_Timers[3]	*LL Pump 1 DV Close Check Timer*	Base	Timer
LLF_RWP1_Runtime[0]	*LL Pump 1 Runtime Hours*	Base	Counter
LLF_RWP1_Runtime[1]	*LL Pump 1 Runtime Tenths of Hours*	Base	Counter
LLF_RWP2_PF_YOK	*LL Pump 2 Available*	Base	Boolean
LLF_RWP2_PB_YRM	*LL Pump 2 Select Remote Manual*	Base	Boolean
LLF_RWP2_PB_YRA	*LL Pump 2 Select Remote Auto*	Base	Boolean
LLF_RWP2_PF_ZRM	*LL Pump 2 Remote Manual Mode*	Base	Boolean
LLF_RWP2_PF_ZRA	*LL Pump 2 Remote Auto Mode*	Base	Boolean
LLF_RWP2_Timers[0]	*LL Pump 2 Start Check Timer*	Base	Timer
LLF_RWP2_Timers[1]	*LL Pump 2 Stop Check Timer*	Base	Timer
LLF_RWP2_Timers[2]	*LL Pump 2 DV Open Check Timer*	Base	Timer
LLF_RWP2_Timers[3]	*LL Pump 2 DV Close Check Timer*	Base	Timer
LLF_RWP2_Runtime[0]	*LL Pump 2 Runtime Hours*	Base	Counter
LLF_RWP2_Runtime[1]	*LL Pump 2 Runtime Tenths of Hours*	Base	Counter

Figure 6.16 Sample software points list.

other software documentation is there, as described in Chapter 4. Once the programmer reviews the design documents, then understanding the structure and naming conventions used in the program becomes much simpler; 'getting up to speed' for someone new takes much less time. The end user or customer of the SCADA system will also appreciate the detailed documentation and the organized approach used in the design and implementation of the application software.

The procedure which I have used successfully on many SCADA projects consists of the following broad steps; these steps will be explained and illustrated shortly.

1. Create a spreadsheet workbook in which the sheets contain the field I/O points for the controller
2. Create a second workbook containing the software or virtual program points for the controller
3. Open the controller programming software package, such as A-B RSLogix 5000
4. Enter a few points to establish the format; these points could be motor status input and motor control output, for example

5. Export the database file to a Comma Separated Value (CSV) file which can then be edited
6. Open the original spreadsheet files containing the edited points; see Figures 6.8 and 6.9
7. Copy and paste the entries from the previous two spreadsheets into the exported CSV file, so as to create the definitions for the controller points in the required format
8. Save the updated CSV file
9. Import the modified CSV file containing the new entries into the controller software.

This approach results in a controller program project with a completed database, avoiding having to enter each point using the programming software. By using the copy and paste and other features of the spreadsheet program (i.e. MS Excel), considerable time can be saved in generating the database for the controller application.

Now the more detailed procedure will be presented, with some illustrations to show the intermediate results of the procedure. The steps have been expanded here to make things a little easier to follow and to provide some illustrations of the results of the steps:

1. Following the examples given in Section 6.3, create a spreadsheet workbook with sufficient sheets to accommodate all of the field I/O points; populate the spreadsheet pages with the signal tagnames, and their descriptions and addresses for each point in the controller.
2. Again using the methods outlined previously, create another spreadsheet workbook with all of the program or software points identified for this controller; in the beginning, there may not be many points, as these points can only be created as the program logic is developed, and hence, the required software points are identified.
3. Save these two spreadsheet files as XLS format as the field I/O points lists and the software/program points lists, using the name of the controller (i.e. Raw Water Station).

Figure 6.17 shows one sheet of the field I/O points list workbook; all of the details for each point have been included, and this will serve as documentation for the hardware points lists.

The number of sheets in the workbook depends upon the number of I/O modules configured in the PPC. The modules, as illustrated in Figure 6.17, should be organized such that any one module does not cross page boundaries so that the sheets are more readable.

Additional sheets may be added to provide configuration information, such as the layout of the modules, and the number of each point type in the PPC. This will be illustrated later in Section 6.3.6.

These two spreadsheet files now contain the initial database points for the project and should be printed out for reference. The next steps require entering these points into the programming software.

4. Start the programming software for the controller, such as the A-B RSLogix 5000 software; create two or three points of any type in the programming software to establish the structure of points; for example, open the Controller Tags database and enter a few sample points to establish the format.
5. From the programming software, export the database points to a CSV file.
6. Open the CSV file in the spreadsheet program (i.e. MS Excel) to view the entries.
7. Take note of the key information columns that are required for defining the database points for the application program (e.g. Type, Name, Description, Data type, Specifier).
8. Save this as a CSV file using the name of the controller.

Water Treatment Plant **Raw Water Pumping Station**		*Controller Name:* *Network IP:*		RWPSTN 192.168.205.141		
Tagname	Point Description	Point Type	Hardware Address	PLC Module Prod. Code	Tag Reference	Terminal Numbers
LLF_RWP2-00-SRN	*LL Pump 2 Running Status*	DI	I:7/0	1746-IA8	GB-020	1120-07
LLF_RWP2-00-SCM	*LL Pump 2 Remote Mode*	DI	I:7/1	1746-IA8	HI-020	1120-10
LLF_RWP2-00-AGF	*LL Pump 2 General Fault*	DI	I:7/2	1746-IA8	XS-020	1120-11
LLF_RWP2-PS-APL	*LL Pump 2 Low Pressure Alarm*	DI	I:7/3	1746-IA8	PSL-021	1120-12
LLF_RWP2-PS-APH	*LL Pump 2 High Pressure Alarm*	DI	I:7/4	1746-IA8	PSH-022	1120-13
LLF_RWP2-DV-SCM	*LL Pump 2 DV Remote Mode*	DI	I:7/5	1746-IA8	HI-024	1120-14
LLF_RWP2-DV-SCD	*LL Pump 2 DV Close Status*	DI	I:7/6	1746-IA8	ZSC-024	1120-15
LLF_RWP2-DV-SOD	*LL Pump 2 DV Open Status*	DI	I:7/7	1746-IA8	ZSO-024	1120-16
LLF_RWP3-00-SRN	*LL Pump 3 Running Status*	DI	O:4/0	1746-OA8	GB-060	1130-00
LLF_RWP3-00-SCM	*LL Pump 3 Remote Mode*	DI	O:4/1	1746-OA8	HI-060	1130-01
LLF_RWP3-00-AGF	*LL Pump 3 General Fault*	DI	O:4/2	1746-OA8	XS-060	1130-02
LLF_RWP3-PS-APL	*LL Pump 3 Low Pressure Alarm*	DI	O:4/3	1746-OA8	PSL-061	1130-03
LLF_RWP3-PS-APH	*LL Pump 3 High Pressure Alarm*	DI	O:4/4	1746-OA8	PSH-062	1130-04
LLF_RWP3-DV-SCM	*LL Pump 3 DV Remote Mode*	DI	O:4/5	1746-OA8	HI-064	1130-05
LLF_RWP3-DV-SCD	*LL Pump 3 DV Close Status*	DI	O:4/6	1746-OA8	ZSC-064	1130-06
LLF_RWP3-DV-SOD	*LL Pump 3 DV Open Status*	DI	O:4/7	1746-OA8	ZSO-064	1130-07

Figure 6.17 Sample field I/O points list sheet.

TYPE	SCOPE	NAME	DESCRIPTION	DATATYPE	SPECIFIER
TAG		Local:0:C		AB:1756_DI_DC_Diag:C:0	
TAG		Local:0:I		AB:1756_DI_DC_Diag:I:0	
TAG		Local:1:C		AB:1756_DO_DC_Diag:C:0	
TAG		Local:1:I		AB:1756_DO_DC_Diag:I:0	
TAG		Local:1:O		AB:1756_DO:O:0	
TAG		Local:3:C		AB:1756_AI6_Float:C:0	
TAG		Local:3:I		AB:1756_AI6_Float:I:0	
TAG		Local:4:C		AB:1756_AO6_Float:C:0	
TAG		Local:4:I		AB:1756_AO6_Float:I:0	
TAG		Local:4:O		AB:1756_AO6_Float:O:0	
TAG		LLF_RWP2_PB_YRA	Select Remote Auto Mode of Control	BOOL	
TAG		LLF_RWP2_PB_YRM	Select Remote Manual Mode of Control	BOOL	
ALIAS		LLF_RWP2-00-SRN LL	Pump 2 Running Status	BOOL	Local:0:I.Data.0
ALIAS		LLF_RWP1-DV-QZI	LLP1 Discharge Valve Position	REAL	Local:3:I.Ch0Data

Figure 6.18 Sample exported database.

In Figure 6.18, there is an illustration of an export operation containing some representative points.

At this point, we have a file that contains the information required for each controller database point organized under the headings that the software requires. In this example, one can see the definitions for the names (tagnames) and the field or alias references. The next steps involve populating the CSV file with the rest of the points for the project.

9. Open both the XLS file and the CSV file so that entries from one can be copied and pasted into the other.
10. Copy the tagnames and descriptions from the XLS file sheets into the corresponding columns of the CSV file; the latter will exist as a single long sheet that continues to grow and become longer as entries are made.
11. Complete the common fields such as 'Type'; using the Copy and Paste function.
12. Ensure that the entries which represent field I/O points have the type 'Alias' designation, and that the software points have the type 'Tag' designation; note that the field I/O points all have a specifier field, which is the hardware address in the controller modules.
13. If desired, this CSV file can be sorted either alphabetically by tagname or by tag type (alias/field or tag/software).
14. Save this file as a CSV file using the controller name.

In Figure 6.19, there is an illustration of the updated CSV files showing additional points. In this example, not all of the points have been added, only a subset to illustrate the method for creating the file to be imported back into the software.

The CSV file should now contain all of the field I/O points and some of the software points for the application program. Note the format for timers and counters, as these are examples of software and support type points. The tagnames define arrays with the number of elements included.

The next step requires importing this information into the controller programming software.

15. From the programming software, select the Import feature to read in the CSV file that has just been created; import all of the entries to build the project database.
16. Verify that the controller database now contains all of the points defined in the XLS spreadsheet files, both field and software types.
17. Save the controller program with the database points.

TYPE	SCOPE	NAME	DESCRIPTION	DATATYPE	SPECIFIER
TAG		LLF_RWP2_PB_YRA	Select Remote Auto Mode of Control	BOOL	
TAG		LLF_RWP2_PB_YRM	Select Remote Manual Mode of Control	BOOL	
ALIAS		LLF_RWP2-00-SRN	LL Pump 2 Running Status	BOOL	Local:0:I.Data.0
ALIAS		LLF_RWP1-DV-QZI	LLP1 Discharge Valve Position	REAL	Local:3:I.Ch0Data
TAG		LLF_RWP2_PB_YRA	Select Remote Auto Mode	BOOL	
TAG		LLF_RWP2_PB_YRM	Select Remote Manual Mode	BOOL	
ALIAS		LLF_RWP2-00-SRN	LL Pump 2 Running Status	BOOL	Local:0:I.Data.0
ALIAS		LLF_RWP1-DV-QZI	LLP1 Discharge Valve Position	REAL	Local:3:I.Ch0Data
TAG		LLF_RWP1_PF_YOK	LL Pump 1 Available	BOOL	
TAG		LLF_RWP1_PB_YRM	LL Pump 1 Select Remote Manual	BOOL	
TAG		LLF_RWP1_PB_YRA	LL Pump 1 Select Remote Auto	BOOL	
TAG		LLF_RWP1_PF_ZRM	LL Pump 1 Remote Manual Mode	BOOL	
TAG		LLF_RWP1_PF_ZRA	LL Pump 1 Remote Auto Mode	BOOL	
ALIAS		LLF_RWP2-00-SRN	LL Pump 2 Running Status	BOOL	Local:0:I.Data.0
ALIAS		LLF_RWP2-00-SCM	LL Pump 2 Remote Mode	BOOL	Local:0:I.Data.1
ALIAS		LLF_RWP2-00-AGF	LL Pump 2 General Fault	BOOL	Local:0:I.Data.2
ALIAS		LLF_RWP2-PS-APL	LL Pump 2 Low Pressure Alarm	BOOL	Local:0:I.Data.3
ALIAS		LLF_RWP2-PS-APH	LL Pump 2 High Pressure Alarm	BOOL	Local:0:I.Data.4
ALIAS		LLF_RWP2-DV-SCM	LL Pump 2 DV Remote Mode	BOOL	Local:0:I.Data.5
ALIAS		LLF_RWP2-DV-SCD	LL Pump 2 DV Close Status	BOOL	Local:0:I.Data.6
ALIAS		LLF_RWP2-DV-SOD	LL Pump 2 DV Open Status	BOOL	Local:0:I.Data.7
ALIAS		LLF_RWP2_DM_DRN	LL Pump 2 Motor Run Control	BOOL	Local:1:O.Data.0
ALIAS		LLF_RWP2_DV_DON	LL Pump 2 Disch. Valve Open	BOOL	Local:1:O.Data.1
ALIAS		LLF_RWP2_DV_DCE	LL Pump 2 Disch. Valve Close	BOOL	Local:1:O.Data.2
ALIAS		LLF_RWP1-DV-QZI	LLP1 Discharge Valve Position	REAL	Local:3:I.Ch0Data
ALIAS		LLF_RWP2-DV-QZI	LLP2 Discharge Valve Position	REAL	Local:3:I.Ch1Data
ALIAS		LLF_RWP3-DV-QZI	LLP3 Discharge Valve Position	REAL	Local:3:I.Ch2Data
ALIAS		LLF_RWP1-DV-KSP	LLP1 Discharge Valve Setpoint	REAL	Local:4:O.Ch0Data
ALIAS		LLF_RWP2-DV-KSP	LLP2 Discharge Valve Setpoint	REAL	Local:3:O.Ch1Data
ALIAS		LLF_RWP3-DV-KSP	LLP3 Discharge Valve Setpoint	REAL	Local:3:O.Ch2Data
TAG		LLF_RWP1_Timers[4]	LL Pump 1 Check Timers	TIMER	
TAG		LLF_RWP1_Runtime[2]	LL Pump 1 Runtime	COUNTER	

Figure 6.19 Sample database with added entries, ready for import.

As the program logic is developed, the additional software points should be added to the XLS file containing the software or internal points. This spreadsheet will become the final documentation for the software points.

For documentation purposes, each of the spreadsheets can be printed on standard sized paper, in landscape format. These spreadsheets, organized by controller, can then be kept as database documentation.

6.3.6 Sample Controller Database

Consider a SCADA project I completed for a customer, which used GE Fanuc PLCs with multiple racks. The racks chosen each had 10 slots, including one slot allocated to the processor module; this left nine slots in the first rack for field I/O modules. The total number of field I/O points was:

- 67 DI
- 44 DO
- 19 AI
- 6 AO

Determining the number of each module type resulted in the following counts:

- 7 DI modules with 16 points each
- 5 DO module with 8 points each
- 5 AI modules with 4 points each
- 4 AO modules with 2 points each

Some of the discrete output modules had 8 points while others had 16 points; the former used isolated outputs while the latter used groups of four outputs with a common line. While other module types can have 8 analog input or output points, this project used the point count shown above.

Assigning the various points to the modules resulted in three racks, with 29 slots available for I/O modules. The final configuration of the racks was as shown in Figure 6.20.

WATER TREATMENT PLANT
Plant PLC Points List

ACME WATER TREATMENT PLANT - MAIN PLC

	Slot	Module	Description
Rack 0 :	1	CPU364	*CPU Processor*
	2	ALG221	*Analog Input Card - 4 Point*
	3	ALG221	*Analog Input Card - 4 Point*
	4	ALG221	*Analog Input Card - 4 Point*
	5	ALG221	*Analog Input Card - 4 Point*
	6	ALG221	*Analog Input Card - 4 Point*
	7	ALG391	*Analog Output Card - 2 Point*
	8	ALG391	*Analog Output Card - 2 Point*
	9	ALG391	*Analog Output Card - 2 Point*
	10	ALG391	*Analog Output Card - 2 Point*
Rack 1 :	1	MDL240	*Digital Input Card - 16 Point*
	2	MDL240	*Digital Input Card - 16 Point*
	3	MDL240	*Digital Input Card - 16 Point*
	4	MDL240	*Digital Input Card - 16 Point*
	5	MDL240	*Digital Input Card - 16 Point*
	6	MDL240	*Digital Input Card - 16 Point*
	7	MDL240	*Digital Input Card - 16 Point*
	8		
	9		
	10		
Rack 2 :	1	MDL930	*Digital Output Card - 8 Point*
	2	MDL930	*Digital Output Card - 8 Point*
	3	MDL930	*Digital Output Card - 8 Point*
	4	MDL930	*Digital Output Card - 8 Point*
	5	MDL930	*Digital Output Card - 8 Point*
	6	MDL940	*Digital Output Card - 16 Point*
	7	MDL940	*Digital Output Card - 16 Point*
	8		
	9		
	10		

Figure 6.20 PLC rack configuration.

WATER TREATMENT PLANT Plant PLC Points List			

ACME WATER TREATMENT PLANT - MAIN PLC

WATER TREATMENT PLANT

GE Fanuc PLC I/O Points Lists

		Used	Spare	Total
Discrete Inputs	:	67	45	112
Discrete Outputs	:	44	28	72
Analog Inputs	:	19	1	20
Analog Outputs	:	6	2	8

Figure 6.21 PLC cover sheet.

A cover sheet for the points lists is shown in Figure 6.21. In this illustration, the total number of points used, the total number of points available and the spare points are identified. This sheet would serve well as the cover, since it contains summary information about the point count, which is useful for future expansion of a project.

Figure 6.22 illustrates one sheet from the workbook showing the assignment of some of the discrete input points in the rack. Note that there is a column for identifying the rack and slot number, as well as a column for the ISA tag cross reference.

6.4 Developing the SCADA Workstation Database

If the SOW is intended to serve as a window into all processes of the system, then all information from all of the PPCs must be available to these SCADA workstations. Through process graphic displays and other user interfaces, the user can monitor and interact with all of the system operations, from one central location. The SOWs are typically located in a group within a central control room for the system.

All of the information contained in the process and historical databases is obtained via the communication network, which reaches out to every PPC in the system. From this central location, operations and supervisory staff can monitor every

ACME WATER TREATMENT PLANT

Plant PLC Points List

Tagname	Point Description	Point Type	Software Address	Hardware Address	PLC Module Prod. Code	Reference	Comment
		DI		%I0065	IC693MDL240		***Rack 1, Slot 5***
WTP_CLA1_00_SCM	*Clarifier Mixer Auto Mode*	DI	%M1061	%I0066	IC693MDL240	SS704	
WTP_CLA1_00_SRN	*Clarifier Mixer Running Status*	DI	%M1062	%I0067	IC693MDL240	VFD705	
WTP_CLA1_00_AGF	*Clarifier Mixer General Fault*	DI	%M1063	%I0068	IC693MDL240	VFD705	
WTP_CLA1_DM_SCM	*Clarifier Scraper Drive Auto Mode*	DI	%M1064	%I0069	IC693MDL240	SS707	
WTP_CLA1_DM_SRN	*Clarifier Scraper Drive Running Status*	DI	%M1065	%I0070	IC693MDL240		
WTP_CLA1_DM_AHH	*Clarifier Scraper Drive High Torque Alarm*	DI	%M1066	%I0071	IC693MDL240		
HLF_HLP1_00_SCM	*High Lift Pump 1 Auto Mode*	DI	%M1071	%I0072	IC693MDL240	SS710	
HLF_HLP1_00_SRN	*High Lift Pump 1 Running Status*	DI	%M1072	%I0073	IC693MDL240	R711	
HLF_HLP1_00_AGF	*High Lift Pump 1 General Fault*	DI	%M1073	%I0074	IC693MDL240	R712	
HLF_HLP2_00_SCM	*High Lift Pump 2 Auto Mode*	DI	%M1081	%I0075	IC693MDL240	SS713	
HLF_HLP2_00_SRN	*High Lift Pump 2 Running Status*	DI	%M1082	%I0076	IC693MDL240	R714	
HLF_HLP2_00_AGF	*High Lift Pump 2 General Fault*	DI	%M1083	%I0077	IC693MDL240	R715	
HLF_HLP3_00_SCM	*High Lift Pump 3 Auto Mode*	DI	%M1091	%I0078	IC693MDL240	SS716	
HLF_HLP3_00_SRN	*High Lift Pump 3 Running Status*	DI	%M1092	%I0079	IC693MDL240	R717	
HLF_HLP3_00_AGF	*High Lift Pump 3 General Fault*	DI	%M1093	%I0080	IC693MDL240	R718	
CHM_AST1_LE_ALL	*Alum Tank Low Level Alarm*	DI	%M1529	%I0081	IC693MDL240	LIT1009	***Rack 1, Slot 6***
CHM_AST1_LT_ALR	*Alum Tank Level - LOE*	DI	%M1530	%I0082	IC693MDL240	LIT1009	
CHM_AFP1_00_SCM	*Alum Feed Pump 1 Auto Mode*	DI	%M1121	%I0083	IC693MDL240	SS727	
CHM_AFP1_00_SRN	*Alum Feed Pump 1 Running Status*	DI	%M1122	%I0084	IC693MDL240	R728	
CHM_AFP2_00_SCM	*Alum Feed Pump 2 Auto Mode*	DI	%M1131	%I0085	IC693MDL240	SS729	
CHM_AFP2_00_SRN	*Alum Feed Pump 2 Running Status*	DI	%M1132	%I0086	IC693MDL240	R730	
CHM_PFP1_00_SRN	*Polymer Feed Pump 1 Auto Mode*	DI	%M1111	%I0087	IC693MDL240	SS731	
CHM_PFP1_00_AGF	*Polymer Feed Pump 1 Running Status*	DI	%M1112	%I0088	IC693MDL240	R732	
CHM_PFS1_MX_SCM	*Polymer Mixer Auto Mode*	DI	%M1118	%I0089	IC693MDL240	SS733	
CHM_PFS1_MX_SRN	*Polymer Mixer Running Status*	DI	%M1119	%I0090	IC693MDL240	R734	
		DI		%I0091	IC693MDL240		
		DI		%I0092	IC693MDL240		
		DI		%I0093	IC693MDL240		
		DI		%I0094	IC693MDL240		
		DI		%I0095	IC693MDL240		
		DI		%I0096	IC693MDL240		

Figure 6.22 Sample field I/O sheet from points list.

aspect of the process control system, and effect control actions and modify operating parameters for each process area. The workstation database, whether it is located in one of the operator workstations or resides on a separate server workstation, contains the combination of databases from all of the field or process controllers (PPCs).

6.4.1 Controller Related Data Points

The creation of the database for the operations workstations requires combining all of the database information for all of the PPCs in the SOW database. Procedures described in the previous section explained how data points from spreadsheets can be transferred into new spreadsheets for the SOWs. This new spreadsheet, in the form of a CSV file, is populated with all of the data points from the various PPC application spreadsheets. The standard method of Copy and Paste can be used to quickly build the new spreadsheets which include all points for all PPCs.

As described for the PPC database development, the programmer must first create one of each data type in the SOW application software. The operator workstation software, such as Wonderware InTouch, includes an application in which the database can be created. For the SOW, there are typically three basic data types, representing the different types of data being handled:

- I/O Discrete – input and output discrete points
- I/O Integer – input and output analog points, typically 32-bit double integers
- I/O Real – input and output analog points, 32-bit floating point

By entering one of each of these data types with tagname and description, an initial database can be created which establishes the format for all entries. As described for the PPC software, this initial database can be exported into a CSV file, which can then be manipulated in Excel. All of the required fields are represented by headings, so the programmer can identify what additional information must be entered for each point.

One key difference between the SOW database and the PPC database is that the SOW database requires a reference to the specific PPC or controller, as the I/O communication driver needs to know how to access each point from each controller. For example, consider a discrete input signal from the Low Lift PPC: LLF_RWP2_00_SRN, which represents the run status of raw water pump 2 in the low lift PPC. In addition to the data type, tagname and description for this point, the SOW database also requires a reference or Node Name for the PPC from which the data is derived. The Wonderware InTouch software uses the term 'Access Name' to refer to the specific PPC.

Using the water treatment plant example, there might be five controllers whose first tagname fragments would appear as follows:

- LLF Low Lift or Raw Water Station
- PRE Pretreatment
- FLT Filtration
- HLF High Lift or Treated Water Station
- CHM Chemical Injection Systems

:IOAccess	Application	Topic
RPS	DASABCIP	RPS_PPC
Galaxy	\\NA\NA	NA
CLX2	DASABCIP	Pumpstation

:IODisc	Group	Comment	OffMsg	OnMsg	AccessName	ItemName
RPS_BSP1_DV_SOD	$System	RPS Pump 1 DV open status	NotOpen	Open	RPS	RPS_BSP1_DV_SOD
RPS_BSP1_DV_SCD	$System	RPS Pump 1 DV Close status	NotClosed	Closed	RPS	RPS_BSP1_DV_SCD
RPS_BSP1_00_DRN	$System	RPS Pump 1 Run Control	Stop	Run	RPS	RPS_BSP1_00_DRN
RPS_BSP1_SS_SCM	$System	RPS Pump 1 Control Mode	Stopped	Running	RPS	RPS_BSP1_SS_SCM

Figure 6.23 Sample discrete point export.

:IOAccess	Application	Topic
RPS	DASABCIP	RPS_PPC
Galaxy	\\NA\NA	NA
CLX2	DASABCIP	Pumpstation

:IOInt	Group	Comment	EngUnits	MinEU	MaxEU	Min Raw	MaxRaw	AccessName	ItemName
RPS_RES1_01_VRT	$System	RPS Pump 1 Run time Hours	%	0	100	0	32767	RPS	RPS_RES1_01_VRT

:IOInt	Group	Comment	EngUnits	MinEU	MaxEU	Min Raw	MaxRaw	AccessName	ItemName
RPS_RES1_01_QLI	$System	RPS Reservoir Level	%	0	5	-32768	32767	RPS	RPS_RES1_01_QLI

Figure 6.24 Sample analog point export.

Using examples from the Wonderware InTouch SCADA software package, a partial database has been created with discrete and analog points. For clarity, some of the fields in the InTouch database entries have been omitted, such as initial value, whether the point is retentive, and the alarm or event nature of the point. The information shown represents the key information for each point.

Figure 6.23 illustrates a sample database export with some of the discrete points entered through the SOW software. This illustration is based upon the Wonderware InTouch software. In these examples, the Access Name if the PPC node, 'RPS' for Remote Pumping Station.

Figure 6.24 illustrates a similar export for analog points. These points include a range for both the raw input value and the scaled engineering value. Typically the scaling of analog values is performed in the PPC application program, but this feature allows the scaling to be done within the SOW database. Again the Access Name provides the link to the I/O server program.

The 'Item Name' is the reference to the specific database point in the controller, while the 'Access Name' identifies the controller; both fields use alphanumeric names. Note in both illustrations that the Item Name is the same as the Tagname; the tagging system within the ControlLogix controller allows the SOW database to reference each point by the tagname, making the database references straightforward.

6.4.2 SCADA Workstation Database Components

As shown in the illustrations previously, the SOW database entries have additional information that the PPC database entries do not have. The PPC database information is that it is important in the SOW database to include the Tagname, Description (e.g. Comment field) and point type (e.g. I/O discrete, I/O real, I/O integer). Now consider the information required for the SOW database entries.

The workstation database includes information for displaying values on the process graphic displays, such as: the text to show when a discrete point is On or Off; whether the input signal represents an alarm type point which is processed with the alarm handling software, or an event type point which is simply recorded; and the engineering scaling information when the scaling is performed in the HMI system.

From the previous illustrations, one can see some additional columns or fields which must be defined for each database point. Following are some of the additional fields that are present with their meaning or purpose; again, these illustrations are based upon the Wonderware InTouch software:

- Group — Alarm group to which point is assigned
- OffMsg — Text to display when point is False
- OnMsg — Text to display when point is True
- Access Name — Node reference to the PPC, used by the server
- ItemName — Reference to the controller database point
- MinRaw — Minimum value for the raw input value
- MaxRaw — Maximum value for the raw input value
- MinEU — Minimum value for the scaled value in engineering units
- MaxEU — Maximum value for the scaled value in engineering units
- EngUnits — Text for engineering units (e.g. %, L/s, kg)

There are other fields required which typically default to set values. For example, if a discrete input point is designated as an alarm signal, then the alarm state (on or off) and the alarm priority of this point are indicated. For an analog input, such as an analog input real, four alarm setpoints can be configured, along with a deadband and rate of change alarm setpoints. Also, the alarm priority for the point is indicated.

In Chapter 10, the structure and organization of the SOW database will be further addressed, along with how the I/O servers must be configured to form the link between the user workstations and the controllers in the system. For the purpose of this chapter, the intent was to illustrate some simple examples of the SOW database.

6.4.3 Procedures for Creating Workstation Spreadsheet Database

Creating the database for the SOW is very similar to the procedure used for the PPCs; however, in the case of the workstations, there are some additional fields of information that must be entered, and the application databases for all of the PPCs must be combined into a single process database.

In Section 6.3.5, a detailed procedure was provided which allows the creation of the final database using the Copy and Paste and other spreadsheet features. Following is a brief summary of this procedure, modified for the workstation database:

1. Open the tag database build software for the SOW package and enter a few points to establish the format
2. Export the database file to a CSV file which can then be edited

3. Copy and Paste the entries from the PPC database points lists created earlier into the exported CSV file, so as to create the database entries for the controller points
4. Using the Copy and Paste feature, fill in the remaining columns for fields such as alarm type, scaling units, etc.
5. Save the updated CSV file
6. Import the modified CSV file containing the new entries into the SOW software package and verify that all fields have been completed.

For documentation purposes, each of the spreadsheets can be printed on standard sized legal paper, in landscape format. Due to the additional fields in the SOW database, I have found using the legal format paper more appropriate. These spreadsheets would then be available for reference.

Figure 6.25 shows a portion of the discrete I/O database for one controller, with some of the fields excluded for clarity.

The spreadsheet is organized by data type for the SOW; all of the discrete I/O points are together, all of the integer I/O points (i.e. 32-bit signed integers) are together, and all of the real I/O points are likewise grouped together. In this example, only a portion of the discrete I/O points has been shown for one controller (RPS). The other two major type groups would also be listed together with the required fields for those data types.

In summary, the spreadsheets for the SOW database would consist of all of the controller points from the individual PPC points lists integrated into one large database file. It is the node reference, or in this case, the Access Name, that indicates which of the PPCs is associated with each of the database entries. The I/O server program which communicates between the SOW and all of the PPCs, uses the node reference name, together with I/O server configuration parameters, to retrieve data from the controllers and to send out command data to the controllers.

In section 10.5, the configuration of the I/O server is explained in a conceptual way, although the illustrations provided in this book are not intended to provide a complete description of how the server programs are configured. For the purpose of this book, the intent is to illustrate the nature of the databases and to provide a brief explanation of the interface between the SOW database and the individual PPC databases.

:IODisc	Group	Comment	OffMsg	OnMsg	AccessNam	ItemUseTa	ItemName
RPS_BSP1_00_SRN	$System	RS Pump 1 Run Status	Stopped	Running	RPS	Yes	RPS_BSP1_00_SRN
RSP_BPS1_SS_SCM	$System	RS Pump 1 Control Mode	Local	Remote	RPS	Yes	RSP_BPS1_SS_SCM
RSP_BPS1_00_AGF	$System	RS Pump 1 General Fault	Normal	Alarm	RPS	Yes	RSP_BPS1_00_AGF
RSP_BSP1_DV_SCD	$System	RS Pump 1 DV Closed Status	NotCls	Closed	RPS	Yes	RSP_BSP1_DV_SCD
RSP_BSP1_DV_SOD	$System	RS Pump 1 DV Open Status	NotOpn	Open	RPS	Yes	RSP_BSP1_DV_SOD
RPS_BSP2_00_SRN	$System	RS Pump 2 Run Status	Stopped	Running	RPS	Yes	RPS_BSP2_00_SRN
RSP_BPS2_SS_SCM	$System	RS Pump 2 Control Mode	Local	Remote	RPS	Yes	RSP_BPS2_SS_SCM
RSP_BPS2_00_AGF	$System	RS Pump 2 General Fault	Normal	Alarm	RPS	Yes	RSP_BPS2_00_AGF
RSP_BSP2_DV_SCD	$System	RS Pump 2 DV Closed Status	NotCls	Closed	RPS	Yes	RSP_BSP2_DV_SCD
RSP_BSP2_DV_SOD	$System	RS Pump 2 DV Open Status	NotOpn	Open	RPS	Yes	RSP_BSP2_DV_SOD
RPS_RES1_LS_ALH	$System	RS Reservoir High Level Alarm	Normal	Alarm	RPS	Yes	RPS_RES1_LS_ALH
RPS_RES1_LS_ALL	$System	RS Reservoir Low Level Alarm	Normal	Alarm	RPS	Yes	RPS_RES1_LS_ALL
RSP_BSP1_00_DRN	$System	RS Pump 1 Run Control	Stop	Run	RPS	Yes	RSP_BSP1_00_DRN
RSP_BSP1_DV_DON	$System	RS Pump 1 DV Open Control	NotOpn	Open	RPS	Yes	RSP_BSP1_DV_DON
RSP_BSP1_DV_DCE	$System	RS Pump 1 DV Close Control	NotCls	Close	RPS	Yes	RSP_BSP1_DV_DCE
RSP_BSP2_00_DRN	$System	RS Pump 2 Run Control	Stop	Run	RPS	Yes	RSP_BSP2_00_DRN
RSP_BSP2_DV_DON	$System	RS Pump 2 DV Open Control	NotOpn	Open	RPS	Yes	RSP_BSP2_DV_DON
RSP_BSP2_DV_DCE	$System	RS Pump 2 DV Close Control	NotCls	Close	RPS	Yes	RSP_BSP2_DV_DCE

Figure 6.25 Sample of discrete points from RPS controller.

7 Process Control Logic Descriptions

Objectives

- Explain the purpose of the process control logic description
- Define the four major sections to the description
- Organize the logic descriptions by process area of the system

Contract documents for SCADA systems typically include a general narrative or description of how the system is intended to operate. It is generally the contractor's responsibility to fill in the details and provide some form of 'shop drawings' which elaborate on the basic system design. All too often, that is as far as the software documentation is taken. Process Control software used in Programmable Process Controllers (PPCs) is very complex and involves multiple operations. All of the design and implementation details of the application software need to be documented.

This chapter describes the Process Control Logic Description (PCLD) which is the primary detailed design narrative for the PPC application software. When developed properly, these PCLDs can serve as the shop drawings for a contract, as they describe in detail exactly how the finished software product will function. As the project progresses and changes are made, the corresponding PCLDs are updated. At the completion of the SCADA project, the PCLDs become an important part of the software documentation, as all of the detailed operations of the application software have now been documented.

7.1 Purpose of PCLDs

The program listing for a typical PPC, if properly commented and documented, should provide a reasonable level of information as to the design and operation of the program. However, many of the design details will not be made clear; the end-user needs to have a complete understanding of the process operations and exactly how the automated program is designed and how it operates. This is the real purpose of the PCLD: to provide a detailed narrative in simple English which describes all operations, modes of controls, alarm handling logic, etc. that are contained within the program logic.

In addition, the PCLD should include special considerations such as interfacing with other PPC programs and/or other subsystems in the overall SCADA system. For example, there may be information in this PPC which is sent to another PPC via the

Designing SCADA Application Software. DOI: http://dx.doi.org/10.1016/B978-0-12-417000-1.00007-5

communication network; or information may be sent via a telephone or fibre optic communication line to some remote site. Hence, the PCLD must provide a complete description of everything pertaining to this PPC application program.

7.2 Structure of PCLDs

The PCLD can be structured in different ways, as long as the important topics or issues are addressed. In this chapter, an example structure is shown, with a completed PCLD illustrating the application of the methodology. The structure of the PCLD shown here, including the major and minor sections, is a suggested format; the important point is to ensure that all aspects of the design are clearly addressed.

The PCLD should be organized into the following four main sections, with the indicated sub-topics; each of these will be described:

System Control Strategy Overview
- Overview of Process Areas
- Summary of Operational Concepts

Facilities and Parameters
- Summary of Equipment
- Key Process Signals
- Control Parameters and Setpoints

Control Logic Descriptions
- General Requirements
- Local Control Mode
- Remote Manual Control Mode
- Remote Automatic Control Mode

Special Considerations
- Software Interlocks
- Hardwired Interlocks
- Failures and Alarms
- Software Interfacing

7.3 System Control Strategy Overview

This is the overview or summary of the control aspects of the PCLD. There would be one such PCLD for each major area of a system. This overview provides a system-level explanation of the purpose and application of this subsystem, without any programming details. It provides the reader with an introductory understanding of the logic description to follow.

7.3.1 Overview of Process Areas

The specific process area being described is identified, which would be one part of the complete system. The purpose and function of this system are described in general terms, conveying the principles of the process operations. If a process area involves more than one process, such as water filtration and filter backwashing, then a separate logic description is required for each one. This is the 'big picture' information, to introduce the reader to the processes involved and the general flow within the SCADA system.

For example, a process overview for a filtration system would explain that there are 'x' number of filters, each of which can operate independently of the others. Each filter includes a flowrate control valve which is to be controlled by the automatic control program in the PPC. For backwashing and cleaning of the filter, an automatic sequence of operations in the PPC control program will step the filter through a series of operations so as to clean the filter media bed.

7.3.2 Summary of Operational Concepts

Having introduced the purpose and application of the PCLD, this section highlights the major process control operations, again without involving any programming detail. This section is also intended to serve as a high-level introduction to the remaining document.

Again consider the filtration area of a water treatment plant; this has two major operations, filtration and backwashing. The general operation of the filter would be described in terms of what is being controlled (i.e. flow control valve) and what the operator input is (i.e. flowrate setpoint). The flowrate setpoint would come from the operator workstation, and the automatic control program would then adjust the flowrate control valve using a PID control loop, so as to maintain the flowrate entered by the operator.

The backwash sequence would also be identified as a series of steps to clean the filter. This series of steps would be controlled by the automatic program, such that valves are opened and closed, and the backwash pump is started and stopped, with its flowrate being controlled by the program also. The details of the backwash sequence would be explained in the detailed 'Control Logic Description', but in this section, only the basic concepts would be presented.

7.4 Facilities and Parameters

This section introduces and describes the major equipment and the setpoints or parameters involved in the process. The equipment and the control aspects are therefore introduced.

7.4.1 Summary of Equipment

The major pieces of equipment involved in the described process would be identified. For example, the pumps or motors involved and the major signals being monitored would be included. In a water system, the process area pumps and the main process signals such as tank level, flowrate and pressure would be listed. Note that the previous section introduced the process operation in a general way; this section now identifies the equipment involved with more details.

7.4.2 Key Process Signals

The operation of a PPC application program requires the monitoring of some key field signals; these signals are identified in this section. By reviewing only this section, the reader can understand what data is being monitored for the process operation described herein. Key signals might include a tank level, discharge flowrate, discharge pressure, chlorine residual, etc. While not all of the signals used by the program need to be listed, this section serves to identify the key controlling field signals.

7.4.3 Control Parameters and Setpoints

The execution of the application program in the PPC involves the use of parameters and setpoints which are typically entered by the operations staff through the SCADA Operations Workstations (SOWs). If the control involves the starting and stopping of pumps based upon a key tank level signal, then the setpoints would include the start and stop levels and any other important parameters required of the program. There may be additional parameters such as time allowances for pumps or conveyors to start or stop and low-level lockout levels for pumping from a reservoir or well. Any parameter associated with this process would be identified in this section.

These parameters would be listed by their respective tagnames with a description of each. The database documentation would provide more detailed information about each such parameter or setpoint.

7.5 Control Logic Descriptions

The control logic descriptions contain all of the details on what the program is designed to do and how it is designed to do it. The previous two sections identify the equipment involved and the principal information being processed. This section provides the detailed descriptions for all of the process operations.

In the introduction portion of this section, the modes of control and process modes of operation would be described, with the details in subsequent sections. The overall logic should be described, without involving all of the details at this point. Thus, this introductory part describes the process operations and sets the stage for the detailed descriptions of the program logic.

7.5.1 General Requirements

The detailed description of operations which follows in this section often requires certain conditions to be met. For example, a pump may be locked out if the level of the reservoir from which water is being pumped is below a low level setpoint. The pumps or motors involved must be set to the Remote or Automatic mode via selections made through the SOW.

SCADA systems typically have three modes of control for the equipment: Local, Remote Manual and Remote Automatic. The first mode involves local pushbuttons and switches located near the equipment and does not involve the PPC. The latter two modes require the hardware mode to be Remote, and the software modes are selected via the software and determine how the SCADA system controls the equipment.

Briefly, the modes of control are listed here, with detailed explanations following:

Local	local panel pushbuttons and switches control equipment,
Remote Manual	operator control through graphic displays,
Remote Automatic	automatic control program in PPC.

These general requirements must identify any abnormal or interlock conditions in conceptual form; the details are covered later in Section 7.6. The description contained in this overview provides more details than the process overview but merely sets the stage for the coming details. The intent is to provide progressively more details so that the reader gains a more elaborate understanding of the process and the specifics of how it all works.

7.5.2 Local Control Mode

Local Control serves as a hardwired backup to the SCADA system; process equipment uses a local control panel with pushbuttons, lights and selector switches for operating the equipment. The SCADA system, and in particular the PPC, has no control over the equipment, although the PPC still monitors the status of the equipment and reports it through the SCADA workstations. A selector switch, labelled 'Hand/Off/Auto' or 'Local/Remote', is used to select the local mode of control or to allow the SCADA system to perform all control operations.

To operate the equipment without the use of the SCADA system, the selector switch must be in the Local position, such that the controls on the panel perform all operations. As stated, the PPC still monitors all statuses and reports the information through to the SOWs.

7.5.3 Remote Manual Control Mode

With the local panel's selector switch set to 'Auto' or 'Remote', the SCADA system can perform all control operations; the PPC has control of the equipment. This remote manual mode requires that the internal SCADA software is set by operator control through the SCADA workstation, using the process graphic displays. A virtual software point denoted as 'Manual' must be set in the software. In this mode of control, the operator can use the process graphic displays to start and stop equipment as though he/she were physically at the local control panel; the benefit here is that the operator can control any of the equipment anywhere in the system from the central control room.

Typically the remote manual mode involves the operator clicking on 'Start' and 'Stop' pushbuttons on the displays. The user would click on the pushbuttons to enact the desired action. These actions are transferred to the PPC application program, which in turn implement the desired commands. The details of how this mode of control works are described in this section.

Certain checks would be performed, such as verifying that the pump starts or stops within an allowed time period; this time period would be one of the setpoint parameters. Otherwise, the user is free to control the equipment, although the software interlocks would still be in effect (e.g. low-level reservoir lockout setpoint).

7.5.4 Remote Automatic Control Mode

The local control panel's selector switch must again be set to 'Auto' or 'Remote' so that the SCADA system has control. The internal software mode selected by the operator through the displays must be automatic control, which means that the application program executing in the PPC performs all control operations. A virtual software point denoted as 'Automatic' must be set in the software. In this mode of control, the automatic control program in the PPC performs all operations; this includes any interlock or safety issues for the equipment.

The operator can at any time change and adjust setpoints and other parameters, but the actual control of the equipment is performed entirely by the PPC application program. This is the normal mode of operation, as one of the reasons for the development of SCADA systems was to provide an automatic mode of control which did not require operator attention or intervention. In the event of an alarm condition, the operator will be notified and he/she may decide to take control using the Remote Manual mode.

Considering the process of operating treated water pumps, the pumps would be started and stopped based upon the start and stop level setpoints entered for each of the duty assignments. The user, through the SOW, can change both the level setpoints and the duty assignments. Other setpoints might include a low-level lockout for the water reservoir, from which the water is being pumped.

7.6 Special Considerations

The process control operations described in the previous sections explain how the software functions under normal conditions, including the handling of errors and alarm conditions. There may be some special conditions in the design, which should be identified; that is the purpose of this section.

7.6.1 Software Interlocks

The application program executing in the PPC may have software or virtual points which come from another PPC or from another process in the PPC. If there is no flow then the chemical pumps are inhibited from being operated, either in remote manual or remote automatic modes. If the level in a supply reservoir is too low, based upon a low level setpoint, the pumps will be locked out in the software.

7.6.2 Hardware Interlocks

There are some instances in which a hardwired backup system may be required, such that in the event of a malfunction of the SCADA software, the hardware backup system will stop operate correctly. In a sewage pumping station, for example, a high-level float switch may be used to energize a pump, regardless of whether or not the high level setpoint for the level signal caused the pump to start. Similarly, a low-level float switch may be used to de-energize a pump regardless of whether or not the low level setpoint triggers the pump to stop.

7.6.3 Failures and Alarms

Operation of the equipment under program control may result in an alarm or a fault condition; this section is intended to identify how the application software handles such conditions, including how the software recovers from a failure.

A typical failure is a pump not starting within the allowed time, resulting in a 'Pump Failed To Start' alarm condition. This alarm would be annunciated to the operator through the SOW, and the operator would then take appropriate action. This may involve changing the duties on the pumps to use an alternate pump while the failed pump is checked.

The ultimate goal of any SCADA system is to maintain continuous operations while handling faults and alarms which can occur during system operations.

7.6.4 Software Interfacing

In some situations, data being processed by one PPC may be needed by another PPC; this data can typically be transferred over the SCADA network between the PPCs. One such scenario involves the connection of field signals to a particular PPC to minimize the wire runs, but the signals are actually required by another PPC in some other panel. Another scenario might have a level signal into one PPC, where it is needed, but that same signal is needed in another PPC for a PID control algorithm. This software interfacing section identifies any such data transfer requirements.

All PPCs include some type of 'Message' instruction which can be programmed to read or write data from one PPC to another PPC on the same network. In the example of the level signal, the originating PPC program would write the level signal information to the other PPC where the signal is needed for a control loop.

There could also be some interfacing to other systems (i.e. DCS equipment) or to other levels of computers in the hierarchy. Data collected by one PPC may be passed through the operator workstations and out through a Wide Area Network to another SCADA system. The description here is limited to what data is being collected and in what form it is to be sent. The details of the external transfer are usually covered in the description of the operator and server workstations.

7.6.5 *Additional Considerations*

In the design of any SCADA system, the intention is to allow the system to operate unattended, while the safety of the personnel is accounted for. Some systems may include safety concerns that must be incorporated into the design of the system, whether it requires software interlocks or hardwired controls.

7.7 Sample PLCD for Pump Station

In Appendix 'C', there is a sample PCLD for a High Lift or Treated Water Station of a water treatment plant; this PCLD describes all of the details of operations performed in the PPC application program.

8 User Operations Reference Manual

Objectives

- Explain the purpose of the User Reference Manual
- Describe the structure and content of the reference document
- Describe the contents of each major section of the manual

The User Reference Manual provides a detailed description of everything that the user can perform through the SCADA Operations User Workstations. Each of the available process graphic displays and historical trend displays is described in terms of what information is being presented and what operations and modes of control are available to the user.

8.1 Structure and Purpose of User Reference Manual

The SCADA User Operations Reference Manual is intended to provide all of the details on what the user can do and how things are done through the SCADA Operations Workstations (SOW). Since these workstations use colour graphic displays and various reporting methods, it is important to have a document which organizes all of the system operations into a logical form, while at the same time, explaining the information being presented in each display and explaining what the user can do and how the user can do. Other topics include accessing historical information for trends and reports, logging into the system and security levels, backup and restore procedures and general system troubleshooting operations.

While every SCADA system is unique and varies considerably in the complexity of the displays and in the control actions available, there are some generally accepted features and procedures that can be documented for any SCADA system. Every SCADA system must handle alarms; the operator must respond to a new alarm by acknowledging it; it should be possible to display alarms by category or group, such as all alarms for a specific Programmable Process Controller (PPC) or process area of the system. Control actions like starting and stopping equipment need some form of pop-up display which contains the control buttons, so that this display can be invisible except when control actions are required. Setpoints and control parameters must be accessible in a popup or other form of display, which includes the appropriate security access level. In the event of a system crash or major shutdown, there needs to be established procedures for restarting the system and restoring all of the application software for both the SOWs and the PPCs.

Designing SCADA Application Software. DOI: http://dx.doi.org/10.1016/B978-0-12-417000-1.00008-7

In the following sections, a suggested organization for the user manual is presented; detailed explanations of each section and their contents will be provided.

8.2 System Overview and Introduction

All SCADA systems, regardless of the application, involve a number of different procedures and operations; these should be described in a logical manner so that the user can quickly locate the information desired and then effect control or obtain the desired information. Following is a summary of the major sections that should be included in this document, and each is described in subsequent sections:

- SCADA System Overview
- Process Graphic Displays
- Alarm and Event Processing
- Historical Trend Displays and Reports
- System Maintenance Procedures

8.2.1 SCADA System Overview

This section should describe the SCADA system from an overall perspective, identifying the PPCs and SOWs in use, along with a system architecture figure showing how the system has been configured. The purpose of the SCADA system should be described in terms of what processes are being monitored and controlled and what process areas exist in the system.

For example, using the water treatment plant illustration, this section of the manual would briefly describe what each of the PPCs does within the water treatment processes. Key information in these PPCs might be identified, such as the level in the elevated water tank is used to control the operations of the High Lift Station pumps.

A system architecture illustration would be helpful to show the user how the overall system has been configured. Figure 8.1 herein shows an architecture illustration which includes the major SCADA elements. A brief explanation of each of the elements in the illustration should be provided, so that the user/reader is familiar with the purpose of each part of the system.

8.2.2 Central Control Facilities

The use and operation of the SOWs would be explained with respect to operator and supervisor functions, system control, security levels for performing various operations and so on. The number of workstations available, including any Operator Interface Terminals (OITs) which are located out in the plant, would be identified with respect to their purposes. The OIT is a local panel which is functionally equivalent to a small dedicated SOW. The filtration section of the plant may have an OIT which allows the user to view information only for the filters in the plant.

Peripheral devices, such as printers and backup devices, should be identified and their use explained. For example, the printer might be used to print hard copies of

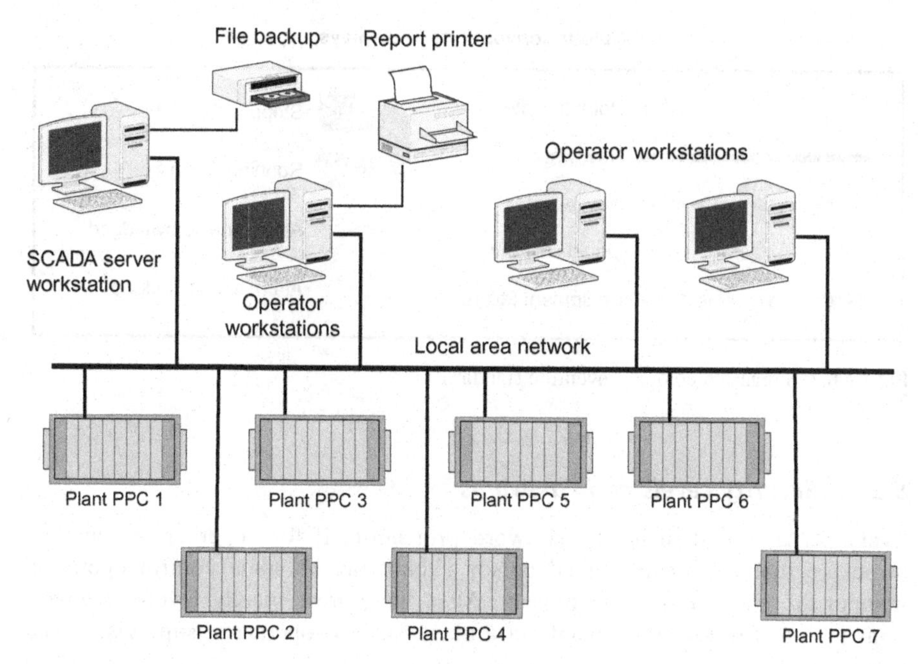

Figure 8.1 Sample SCADA system architecture illustration.

the screen displays and/or to print historical reports. There may also be a modem or other interface for connecting outside of the SCADA system; a brief explanation of the purpose and use of such devices would be covered.

8.2.3 Display Conventions and Navigation

In Chapter 10, the topic of colours used and methods of accessing the various displays is addressed in detail. In this portion of the User Reference Manual, the user/reader should be shown what colour conventions have been chosen and how they are used; explain whether the running status is Red following the electrical convention or Green following the traffic light convention. Likewise, the choice of colours for process lines, chemical lines, alarm conditions, etc. should be explained. A simple summary with the colour choices could be used with each colour being identified as to its application.

Figure 8.2 shows a suggested colour convention that could be used. The conventions used in a system should be documented in a similar manner.

Systems typically have many displays of different types, such as process graphics, trends and alarm/event summaries. The general method of navigating these displays should be explained, leaving the specific details to the subsequent sections of the manual.

Any other general methods or conventions used in the SCADA system should be explained in this section to introduce the user/reader on how the workstations are used in a general sense.

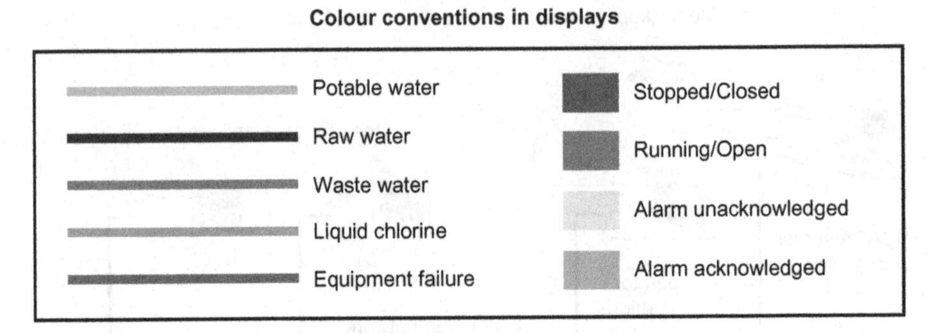

Figure 8.2 Suggested colour convention standard.

8.2.4 Security Levels and Passwords

Every SCADA system needs password protection, if for no other reason than to ensure that any unauthorized person cannot invoke some control operation without the plant staff being present. Also, there are typically levels of operations assigned to different people in a plant, such as operators, supervisors and programmers.

A procedure must exist for logging into and out of the system via the SOWs. Many system designs follow the Microsoft Windows method using a User Name and a Password. When the users and their passwords are configured, usually by the supervisor or programmer, the users are assigned different levels of functions. When a user logs into a workstation, their level of access then applies to everything on that workstation.

The plant operators must be able to view all displays, effect changes to control modes for equipment and operate the equipment under Remote Manual control. They must be able to switch equipment between the Remote Manual and Remote Automatic modes of control. The operator needs to be able to view trend displays and handle alarm conditions in the system.

The supervisors, in addition to the operator-level functions, must also be able to change alarm setpoints and other control setpoint parameters for the system. There may also be system maintenance-type operations that should only be performed by the supervisory personnel.

At the SCADA system level, the software or programmer personnel should have access to everything in the system, as it is their responsibility to maintain the system and to resolve any SCADA issues in the system. Some facilities are large enough that they have their own system programming people; other facilities may have a system programmer person on contract who can be called in when there is a problem.

This security section of the user manual would explain the various levels of security and the operations available at each level. Passwords, of course, would be kept secret and not to be documented anywhere. Screen captures or images of the log-in and log-out procedures would be helpful to the user.

8.3 System Graphic Displays

Process graphic displays contain animation features to convey information to the user of the system. Animation includes colour changes to equipment, visibility of text such as control modes and graphic changes such as levels in tanks and reservoirs. Some displays, like the historical trend display, plot selected values against time to show how process variables change over a selected time period. Still other displays may include summary data, such as runtimes for motors or other equipment.

These sections on the graphic displays should include a screen capture or an image of the screen with the title of the display, an explanation of the displays purpose, what key information is being displayed and what actions are available to the user. Some displays may be information only, such as menus; other displays will include details for a process area, in which case, there will be additional displays like pop-up panels that will have control buttons for invoking actions on the selected equipment. The required details are addressed in each of the following sections, describing the various types of graphic displays that are available in the SCADA system application.

8.3.1 System Overview Displays

Displays which show the entire SCADA system, such as the plant overview, would be described in detail in this section of the manual.

Most SCADA systems include at least one graphic overview display and one main menu display. The overview and/or system type displays should be shown and explained. Typically these system overview displays do not include any control operations or setpoint adjustment functions, so it is sufficient to describe the displays.

In addition to menu displays, the system might have hyperlinks in the overview displays, such that by clicking on certain areas of the overview display, the detailed process graphic display for that area is presented. This is a common method of accessing area displays and should be described and explained in this section.

Text-based menu displays can be very useful as they typically list all of the displays in the system with a button next to each display name; alternatively, instead of a button, the text name of the display could include a hyperlink which when clicked would cause that display to be presented. This section of the user manual should explain the navigation methodology of the SCADA system, so that the user can easily and quickly find whatever display is desired.

Figure 10.10 shows a hierarchy structure which could be used to show the user how the various displays are organized and how to access each display. Whatever the method of navigation is used, it must be consistent and must be explained here in the overview display section of the user manual, so that the user will understand that the navigation is the same throughout.

8.3.2 Alarms and Events Displays

The alarm summary and the event summary displays are tabular in nature and present a chronological list of all alarms or all events in the system. Section 10.4.2 provides the details of these displays.

From the perspective of the user, the manual should explain what options are available for sorting and/or displaying the information in these displays. It might be possible to list only the points that are currently in alarm or only those alarms associated with one process area of the system. So any options available for displaying the information in different forms should be explained.

Acknowledging alarms may require the simple clicking on an 'Acknowledge' button on the screen or may require additional actions by the user/operator. Some systems include an audible alarm which must be silenced; perhaps clicking on the acknowledge alarm button will silence the audible alarm, but additional steps might be required to deal with the actual alarm condition. Whatever steps are needed to acknowledge an alarm must be described so that the user can follow the steps to deal with an alarm correctly.

8.4 Process Graphic Displays

Each of the process areas of the system will have one or more process graphic displays associated with it; in this section, the details of those displays are provided. Most such process displays include control operations and parameter adjustment operations; depending upon how the security levels are defined, some of these operations may require a supervisory or higher level access to perform.

For example, to change the control mode of equipment or to start and stop the equipment, the operator-level access should be sufficient; in this case, clicking on the equipment (i.e. pump and valve) would cause a pop-up display to appear which enables the user to control that equipment. Again, the available operations in each display should be explained.

All of the process graphic displays should be organized into groups, such that there is one section in the user manual for each process area. Each such section would include all of the displays for that process area. Considering the water treatment plant application, this section would be further divided into one section for each of Low Lift, Pretreatment, Filtration, High Lift and Chemical Systems. The descriptions which follow apply to each of the displays in each of the process area groupings of displays.

Following is the general summary of the categories of information that should be included in the user manual for each of the detailed process graphic displays.

Process Overview Description

A description of what the display represents and what process or processes are represented in the display should be explained; also, a reference could be made to the modes of control available for the equipment shown.

Information Display

The details of process operations are summarized, such as the current flowrate and total flow for the station, the equipment statuses (running/stopped/failed), modes of operation and key process signals. A simple point form summary of the information is usually satisfactory.

Equipment Control

The equipment and parameter controls available to the user, based upon the security level, e.g. starting and stopping of field devices, changing control modes and adjusting setpoint parameters. If pop-up control panels are used, then these should be illustrated here; refer to Figure 10.11 for examples of pop-up panels containing control functions.

To illustrate the format being suggested here, I am using an example from a previous project of mine for the Clarifier portion of a water treatment plant. The material between the two double lines is an excerpt from the User Reference Manual for the project.

There may be some operations, such as a backwashing sequence for a filter, that need to be explained step by step. For a filter to be backwashed automatically, the backwash must be started or initiated by the operator; how this is done needs to be explained herein. The filter may have to be placed 'Out of Service' first before a backwash can be initiated. There may be additional states of the filter which should also be described.

Clarifier Operations

Overview

This screen shows the status of the raw water inlet valve, the raw water inlet flowrate and the clarifier effluent turbidity; the status of the clarifier mixer and scraper is shown.

Process Description

The raw water from the low lift station is pumped through the inlet valve and into the clarifier. Polymer and alum are injected at the clarifier to produce floc, which settles to the bottom. The supernatant or clean water passes through to the filter inlet chamber. The clarifier includes a mixer and scraper to aid in the flocculation process and to collect the floc at the bottom of the clarifier.

Information Display

The clarifier is shown with its mixer and scraper, backflush valve and blowdown valve. The mixer and scraper change colour to indicate the current status, i.e. stopped, running or failed. The backflush and blowdown valves operate on command by the operator.

Other current operating information consists of:

- Raw water inlet flowrate, in L/s;
- Clarifier effluent turbidity, in NTU;
- Filter inlet chamber level, in m;
- Current speed of the clarifier scraper and mixer.

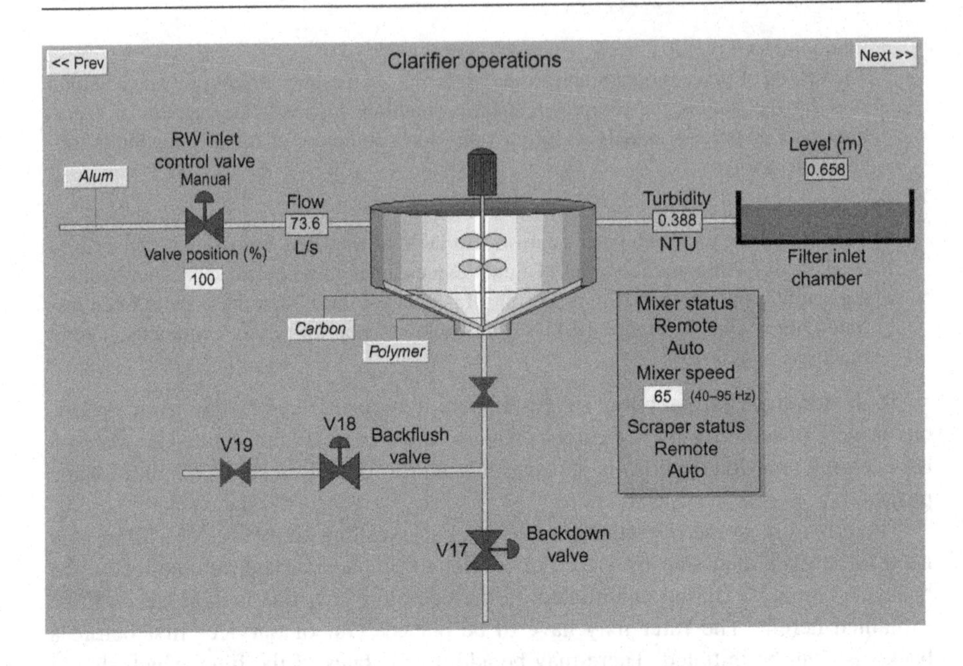

Control Operations

Clarifier Control

Clicking on the 'Controls' pushbutton provides access to the operating modes and start/stop control for the clarifier mixer and scraper. In the 'Manual' mode, the operator can start and stop the mixer and the scraper. In the 'Automatic' mode, both mixer and scraper will operate when there is a minimum flowrate entering the clarifier.

The operator can click on the 'Mixer Speed' field on the main display to change the current operating speed of the mixer: the speed range is from 40 to 95 Hz).

Control Setpoints

The operator can click on the 'Setpoints' pushbutton to access the control setpoints pop-up display; from this pop-up display, the operator can change operating setpoints, as well as start the backflush/blowdown valve cycle.

The operating setpoints consist of:

- Clarifier Blowdown Valve operating or open time, in seconds;
- Clarifier Backflush Valve operating or open time, in seconds;
- Start Delay Time allowed for the mixer and scraper to start when commanded;
- Stop Delay Time for the mixer to continue running after a plant shutdown.

The 'Start' pushbutton on the setpoints pop-up display initiates the following sequence of operations for the valves:

- The Backflush Valve is opened and kept open for the duration setpoint;
- The Backflush Valve is closed, and a delay period must expire before the Blowdown Valve is operated;
- The Blowdown Valve is open and kept open for the duration setpoint;
- The Blowdown Valve is closed.

One water treatment plant had two filters which could be backwashed separately, one at a time. There were three states in which the filter could be In Service and filtering, Out of Service and idle and Backwashing which involved a series of operations normally performed by the PPC program. If desired, the operator could backwash the filter manually by following the steps described in the section on the filters. The actions required to place a filter in service and out of service were described in the user manual, along with the complete set of steps required to backwash a filter.

8.5 Historical Reports and Trend Displays

The previous categories of displays relate to the current operating conditions of the SCADA system; the current levels in tanks and reservoirs, the status of the conveyors and product handling equipment and other ongoing operations. The historical category of displays relate to reports of information that show data values over time. The trend display is the most common and well known of this type of display. But there could be other historical type displays which show summary type data which would be useful to the operator and supervisor of a facility.

8.5.1 Trend Display Operations

The trend displays in a SCADA system all have the same basic format: process values are plotted from left to right with the *Y*-axis showing the value. When trend displays are created and configured, the selected points are shown as a legend with their tagnames and descriptions on the trend display, along with their respective colour, that being the 'pen' colour used to show the value across time.

Most trend display features in SCADA software packages include flexibility in the choice of colours, the number of points to show on one trend, the time base or period for the trend display and details such as background colour and grid lines. In the user manual, each of the trend displays should be shown and described in terms of what the user can do. Perhaps they can double-click on the trend to access a 'chart properties' type window in which the parameters of the trend can be changed. The time period on the *X*-axis can usually be changed, and sometimes the scaling on the *Y*-axis can be changed. The operator can often zoom in to select a portion of the trend, which is then expanded to fill the time scale of the trend display.

All of the configuration and display options should be described, usually in the introduction part of the trend display section, with each of the trend displays being shown separately. Any operations that are applicable to all trends would be described in detail in this section of the user manual.

8.5.2 Historical Reports

A historical display typically contains columns of data, such as the runtimes for all of the motors or pumps in the system. A historical report might include both the current runtimes and the total runtimes since the beginning of the week, month or year.

This historical type display could include the ability to reset the runtimes to zero, although this may be a supervisor-level action.

Another historical type report could show the total flows at various places in a water treatment plant, including the current day's total, the month's total and the year's total. In a materials handling system, this report might show the production counts or totals for each conveyor line. This type of report allows personnel to compare the flows for different time periods and compare the operation of equipment (e.g. each filter in a plant has a flowrate which can sum up to produce a total flow). The actions available, such as resetting the total to zero, would be described in this section.

8.5.3 Exporting Historical Data

If the historical database, which maintains all of the data for the trends and historical reports, uses the Structured Query Language (SQL) format, then exporting data can be relatively simple. There are third-party applications which can extract selected data values from the historical database for a specific time period and then import the data into a spreadsheet for analysis. Once in a spreadsheet, the user has all the power of the spreadsheet program to plot and graph the data, as well as to combine the plant data with other information.

If the SCADA system includes this feature, then a section like this would be needed to explain the steps required to select the desired data records for exporting and to explain how that exported data can be used.

8.6 Special Operating Procedures

The SCADA system is designed to operate continuously without interruption; however, there may be situations in which one or more of the SOWs need to be restarted; hence, there should be well-documented procedures for this so that the shutdown/startup procedure proceeds smoothly with each step happening in the proper order.

8.6.1 Startup and Shutdown Procedures

In the event that the central facility needs to be restarted, there needs to be a standard procedure for a controlled shutdown of the system and startup of the system. The PPCs operate continuously, without any need for restarting, perhaps in the event of a fault in the processor.

The Startup Procedure should be prepared with the assumption that everything is turned off. Then step by step, the equipment is powered up and users can log into the system. Since the server workstation includes the I/O server communication software for connecting with the PPCs over the network, it is important that this software be started before the SCADA operations Human–Machine Interface (HMI) software is started on the workstations. There may be multiple application programs involved which also need to be started in a set order; this sequence must be followed for proper operation of the system.

When the I/O server workstation has initialized, it will be necessary to log into the system, probably at the supervisor level, to start up the various software systems such as the I/O communications and the other HMI applications. When the user workstations have initialized, a 'home' screen should be presented; the user can then log into the system to perform whatever operations are desired.

The Shutdown Procedure should be a controlled shutdown of the workstations. Typically each of the application programs, such as the HMI and I/O server applications, must be stopped first; then the Microsoft Windows can be shut down along with the workstation itself.

Both procedures should be described such that the startup and shutdown operations are clearly laid out and proceeded in a logical manner. An operator should be able to properly start up or shut down a workstation by following these procedures.

8.6.2 System Backup Procedures

Likewise, the data files accumulating on the SCADA server workstation need to be backed up to tape or other storage media on a regular basis, so as to save all of the historically accumulated data from the SCADA system. In some systems, this historical data may be transferred to another server computer for long-term storage; such a transfer would likely be performed on a regular basis, either manually or automatically.

The I/O server workstation maintains the historical database containing values for the selected points over time. It is important to back these files up on a regular basis, whether the backup media is tape, CD or solid-state storage device, such as an external storage device. This section of the user manual must describe the procedure in a step-by-step manner, so that the operator or supervisor can properly save the historical data files.

If any changes have been made to the application software, such as adding points to the process database or modifying process graphic displays, then the application software for the system should also be backed up. Again the choice of storage/backup media is up to the developer and customer.

If proper backup procedures are documented and followed regularly, then in the unfortunate event of a computer problem, one knows that all of the application software and the databases are saved away and can be reloaded once the workstations have been serviced.

9 Guidelines for Controller Application Programming

Objectives

- Organizing the program logic for the controller
- Using various tasks for efficient operation
- Modularizing the code for I/O isolation
- Implementing a design and documentation methodology
- Applying styles in programming

Every Programmable Process Controller (PPC) in a SCADA system executes a control program developed specifically for the process area. This program might consist of a single main routine executing continuously, or a combination of continuous, synchronous and asynchronous tasks, each of which includes multiple programs and routines. This chapter offers guidelines to the design and development of the automatic control programs that execute in these controllers.

9.1 Identifying the Controller Processes

From the Process Control Logic Descriptions (PCLDs), the specific processes required must be identified for each controller. Some operations involve monitoring and collecting information, such as production data or equipment runtimes. Other operations involve closed loop control using the Proportional Integral Derivative (PID) instruction. Still other operations require the control of equipment in Manual and Automatic modes of control, based on the user's selections.

9.1.1 Isolate and Define Processes

The PCLD was developed to identify and describe the operational requirements for the PPC. The PCLDs must be analyzed to isolate specific operations, such as the modes of control for the equipment. By reviewing the details of the descriptions in terms of what monitor and/or control functions are needed, a detailed list of program operations can be determined.

Reviewing the PPC field I/O points list together with the PCLD for the PPC should help to identify the nature of the operations and the specific requirements of the application software. For example, if the PCLD describes control modes for pumps as

Designing SCADA Application Software. DOI: http://dx.doi.org/10.1016/B978-0-12-417000-1.00009-9

Remote Manual and Remote Automatic, and the field I/O points list shows pump run status inputs and pump control outputs, then it can be concluded that there needs to be logic to operate the pumps in two control modes. If the field I/O points list includes a reservoir level analog input signal, and the PCLD describes logic to declare alarms when the current level crosses over a set of level setpoints, then there needs to be logic to compare the level against these setpoints, and set or clear alarm flags accordingly.

After careful review, a specific list of processes or operations can be developed. Initially, this list may simply state the process, without any details; the next step would be to 'flesh out' the details of each process such that there is a conceptual description for each listed process. The software requirements for the process, such as the setpoints and virtual points required, will not be known at this time, but the general or conceptual details of the processes should be established.

9.1.2 Organize Processes for Programming

Whether there is 1 process or 10 processes involved, the listed processes can each be defined in terms of the field I/O signals involved and the estimated software or virtual points required. Using the example earlier of the Manual and Automatic control modes, there would need to be software points for each mode, plus virtual 'pushbuttons' from the host Human–Machine Interface (HMI) facility to set and clear the mode of control; only one of the two modes can be active at any one time, so appropriate interlocking of the software must be included. The development of the actual program code or logic will be addressed later in another section.

The processes can be categorized into groups such as continuous control logic, periodic or cyclic operations and asynchronous or event type operations. Since today's controllers support all of these types of tasks and functions, one can assign the identified processes to the respective group. Data collection, for example, might be assigned to a periodic or cyclic task, which is triggered at regular intervals to collect and save data values. Control of equipment would be part of the continuous or main task of the controller, as these operations are continuous or ongoing.

In reviewing a PCLD for a Low Lift or Raw Water Station at a water treatment plant, the following processes might be listed for programming:

- Travelling screen to operate automatically based on pressure differential on either side of the screen.
- Discharge valves on pumps controlled to maintain setpoint flowrate from station to pretreatment facility.
- Pumps operate in both Manual and Automatic modes; facilities must allow for operator switching modes on each pump.
- Pump and discharge valve must be controlled in sequence such that the pump starts first and the valve opens afterward.
- Duty selection can be made for the two pumps such that either pump can be the Lead or Lag duty unit.
- Water quality is monitored for temperature, turbidity and pH.
- Pump runtimes are accumulated in hours and tenths of hours.
- Flowrate through meter is totalized to produce total flow per day.
- Operating data is transferred to pretreatment PPC for processing.

With this list of processes, the next step is to structure the processes into tasks and programs. The continuous or main task typically contains the majority of the program logic, so this task must be designed with appropriate subroutines and program structure.

9.2 Creating the Application Databases

The application database has been described in other sections of this book, and the detailed method of creating the database has also been described. For controllers with multiple tasks and programs, one must consider the creation of the task/program-specific databases. This section explains how the database is created and maintained for each of the PPC application projects.

9.2.1 Importing Spreadsheet Points Lists

The first step in creating the PPC application database is to create the field I/O points from the spreadsheets created earlier. In section 6.3.5 of this book, a detailed procedure was presented for creating the field I/O points list spreadsheets, and then transferring the entries to the controller project database. This first step establishes the basic hardware points for the application.

Referring to the description in Section 6.3.5, the user must create one or two tagname entries in the controller project and then export this to a Comma Separated Value (CSV) file. The purpose of this step is to establish the format of the project database. This CSV file can be opened in MS Excel, along with the field I/O points lists. Using the copy and paste values feature of Excel, the tagnames, descriptions and addresses can be transferred to the CSV file. This file is then saved, and finally imported back into the controller project, maintaining existing points while adding the new points.

The next step in creating the controller application database is to create the software or virtual points for the program logic.

9.2.2 Defining and Refining Software Points

As the processes are expanded with more details, the required software points will become known. An initial points list of software points can be prepared and imported into the controller project in the same way that the field I/O points were imported. As was illustrated earlier in Chapter 6 regarding database development, spreadsheet lists can be prepared for both the field I/O points and the virtual software points.

As an example, referring to the description in Section 9.1.2, the travelling screen points are as follows:

LLF_MCS1_PF_ZRM	Microstrainer Remote Manual Selected
LLF_MCS1_PF_ZRA	Microstrainer Remote Automatic Selected
LLF_MCS1_PF_XRM	Microstrainer Select Remote Manual Mode
LLF_MCS1_PF_XRA	Microstrainer Select Remote Automatic Mode
LLF_MCS1_BW_XON	Microstrainer Backwash Cycle Active

Similarly, the other equipment identified in the PCLD would have both field I/O hardware points and virtual software points. The hardware points will be established at the start of the programming work, but the software points will develop as the programming proceeds.

It is important to maintain the software points lists as the application software is developed. This ensures that not only does one avoid duplicating tagnames and/or point names, but a record is maintained for all of the software points used in the application program. Once the project has been completed, these software points lists become part of the final software documentation.

9.3 Tasks, Programs and Program Structures

The various processes defined in the PCLDs for a PPC controller must be allocated to one or more programs in the project. Some processes may be best implemented as synchronous or periodic operations, while others may be asynchronous or event type. With today's PPCs, such as the A-B ControlLogix PAC and the GE Fanuc PACSystem controllers, the application program for a single controller can be structured into several tasks, programs, and routines.

Note that the IEC 61131-3 programming standard establishes these types of operations, including continuous, synchronous or periodic and asynchronous or event. The standard establishes an environment which allows for multiple programs with any number of routines; it also allows for multiple tasks, each of which may have a number of programs. Consequently, the application software project can be organized or structured in a variety of ways, depending on the needs of the project.

Each of the PPC controllers identified for a SCADA application will require one or more PCLDs, as each PCLD pertains to one process within the controller. A High Lift or Treated Water controller, for example, involves not just the manual and automatic control of pumps but also the processing of alarms, chemical injection systems, data collection and accumulation, interfacing with other PPCs and the host SCADA workstations, to name a few operations. The typical PPC control program would be structured with one or more tasks and programs for these various processes in the treated water facility of the water treatment plant.

9.3.1 Organize the Controller Project

The traditional PPC program structure utilized a mainline routine and multiple subroutines; each subroutine contained logic to perform specific process operations. The single program executed continuously, with the various subroutines being called to implement the different processes for the application.

Today, PPCs have been enhanced to allow both multiple programs, each with subroutines, and multiple tasks, each of which can have multiple programs. This approach allows much more flexibility in how an application program can be organized and structured.

The tasks are typically any of three basic types: continuous, synchronous or periodic, and asynchronous or interrupt/event. The continuous task is the original program that executes continuously. A cyclic or periodic task is one which is scheduled to execute at set time intervals, such as every 50 ms or every 70 ms; the programmer can set the interval for each of these cyclic tasks, and each task can have multiple programs with subroutines. The interrupt or event task is one that is triggered by some field condition (e.g. discrete input turning on) or by some other identifiable state or condition in the process (e.g. motion control operations reaching a particular position).

Figure 9.1 illustrates the main Continuous task, in which all programs within the task are executed in order in a continuous manner. The order of the program execution is configured by the programmer.

The synchronous and asynchronous tasks are illustrated in Figures 9.2 and 9.3, respectively. Note that the synchronous or periodic task is executed repeatedly at fixed time intervals. The asynchronous or interrupt/event task is executed when a predefined condition occurs, such as a particular input turning on or off.

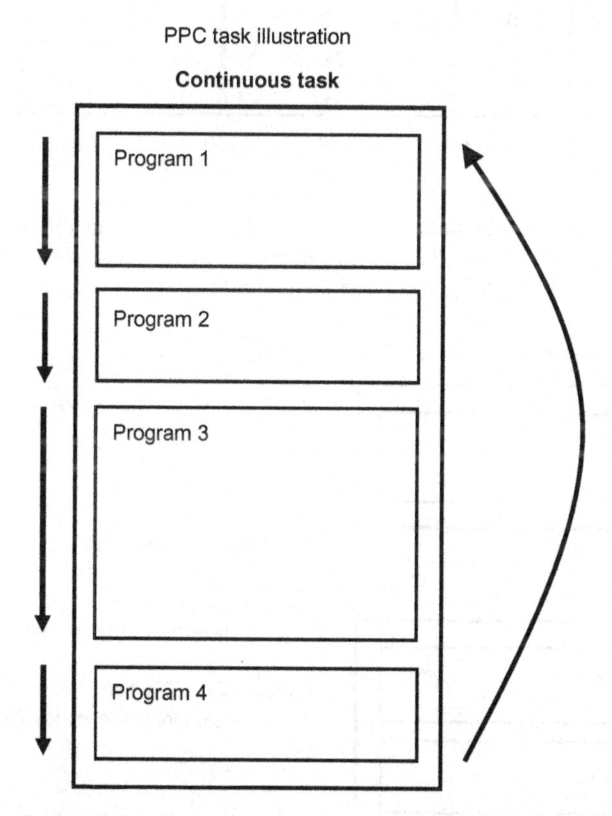

Figure 9.1 Continuous Task – Each program within the task executes once through, and then execution resumes again from the beginning.

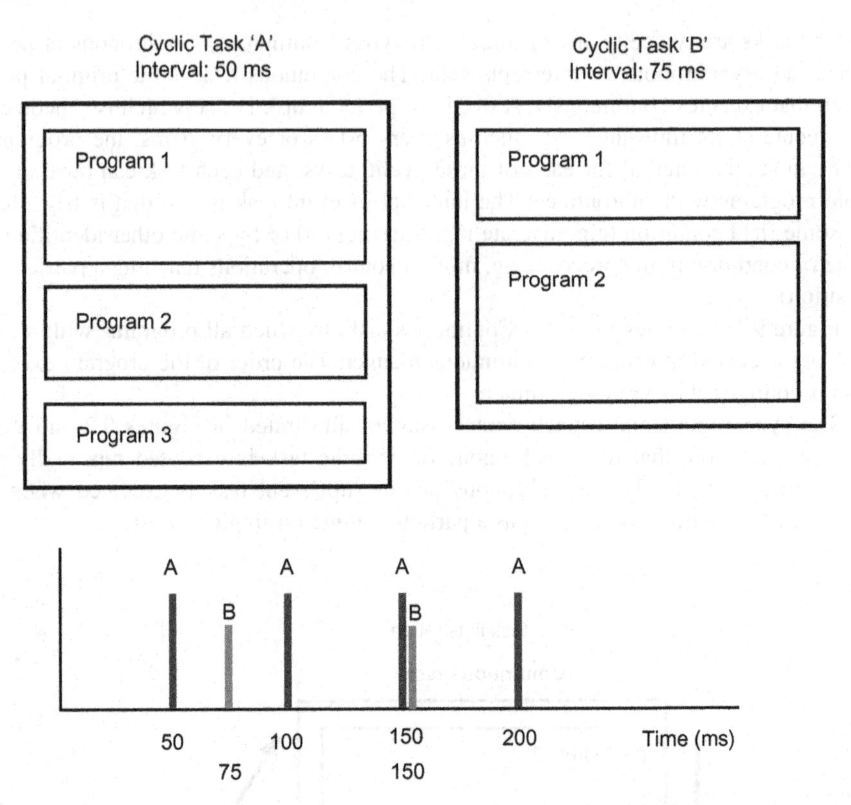

Figure 9.2 Synchronous or Periodic Task – each synchronous task with different scheduled time intervals would execute based on the time interval. When both are scheduled to execute at the same time, the task with the higher priority executes first.

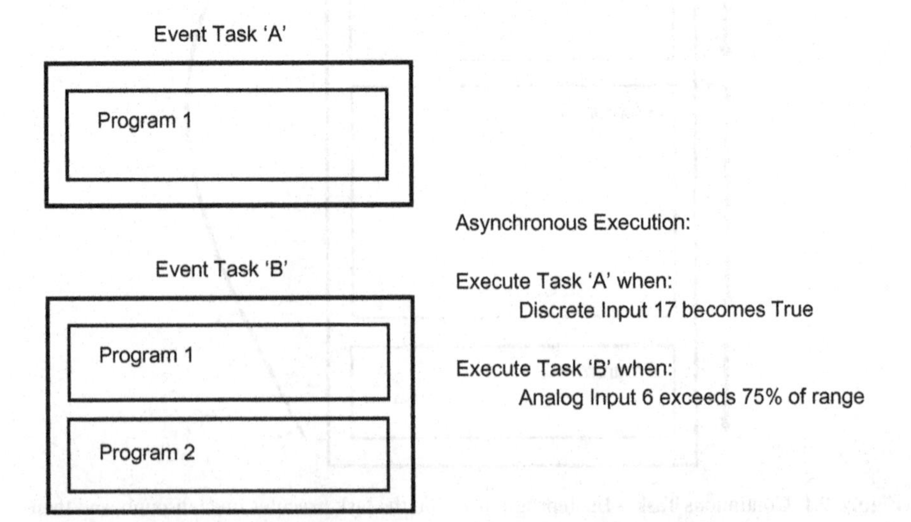

Figure 9.3 Asynchronous or Interrupt/Event Task – each task executes when its respective condition is met.

The A-B ControlLogix PPC allows a total of 32 tasks, each of which can have up to 32 programs. There is only one continuous task, which is also the lowest priority task for the controller; if any of the synchronous or asynchronous tasks is required to execute, then this one continuous task is suspended until the other task has completed one pass or scan through its logic. So, the remaining 31 tasks can be any combination of cyclic and event type tasks. This structure allows for maximum flexibility in program design.

As an example, two synchronous or periodic applications might be a PID control loop and a data collection operation. The PID instruction must execute at fixed time intervals so that the control loop operates correctly, especially since most PID control systems use the 'Integral' function, which is time interval based. Runtime accumulation for motors would be maintained using timers and counters; a cyclic task with at least one program could be used to check the status of the motors and add to the runtimes for those motors. In this way, one could ensure that the status of the motors and hence the time accumulation is being updated at fixed intervals.

A given system may include some critical alarm inputs that require immediate action in the program logic; waiting for the normal scan through the continuous task may not be acceptable. In this case, an event type task could be defined which is triggered when any of these system critical alarms occurs, with the program in the event task responding to the field condition.

So, to best organize a controller project, one must structure all of the required processes into the task and program type that best suits the processes. The continuous task and program(s) can perform the usual continuous processing operations, while the synchronous and asynchronous tasks could be configured for the scheduled and asynchronous needs of the application.

9.3.2 Define the Required Tasks and Programs

In Section 9.1.2, a number of processes were identified related to the Raw Water Pumping Station, such as the operation of the travelling screens and the pumps. Using this example, the task and program structure will be developed.

From the list, the following processes were identified:

- Travelling screen operation
- Raw water pump control
- Duty selection for the pumps
- Data monitoring for water quality and flowrate
- Runtime accumulation for the pumps
- Interfacing with the Pretreatment PPC
- Accumulation of flowrate into total flow.

This application is relatively small, so a single program can be created for most of the logic. A synchronous task and program could be created for the data collection processes to ensure that data values are being sampled at regular fixed time intervals.

A main task with one continuous running program would be created, which included subroutines for each of the processes identified previously. The subroutines for the continuous task and program would be as follows:

- Pump Duty Control
- Pump Control
- Travelling Screen Control
- Current Data Monitoring

The synchronous or periodic task and program would contain subroutines to handle the fixed interval operations; this might execute 5 or 10 times per second:

- Totalling Flow
- Pump Runtime Accumulation
- Pretreatment PPC Interface

9.3.3 Creating the Main Program Structure

The default task and program for the PPC, regardless of the manufacturer, would include a main routine which calls the various subroutines to implement the various processes. An illustration of the program structure is shown in Figure 9.4.

The mainline routine, 'Raw_Water_Station', would call the various subroutines shown to execute the process-specific logic. This main routine may also contain some

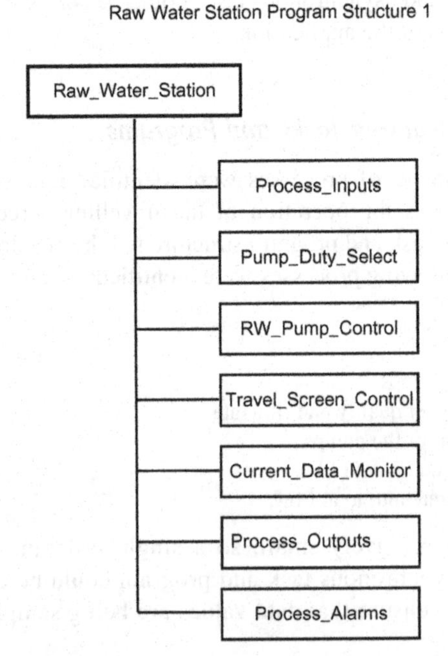

Figure 9.4 Continuous program structure.

general logic associated with the entire application. Alternatively, another routine for 'General_Functions' could be added to handle such general logic requirements.

A brief explanation to each of the subroutines shown in Figure 9.4 is described here.

Process_Inputs

Discrete inputs sometimes require debouncing, such as float switches which may toggle between on and off before stabilizing in one of the two states; timers could be used to stabilize the input before storing the 'debounced' states of such inputs.

By transferring hardware inputs to virtual software inputs, the field signals can be isolated from the internal software so that testing of the internal logic may be done without impacting the field signals. The programmer can manipulate the internal software points to verify the operation of the program logic, thus avoiding interacting with the actual field signals wired to the PPC.

Pump_Duty_Select

Pumping stations, or any other application that involves multiple devices, may require the assignment of duties for operation. For example, with two pumps, one would be designated the Lead duty while the other one would be designated the Lag duty. The user, through the SCADA Operations Workstations (SOW), can assign these duties through the user interface, which would set virtual duty points in the software.

An application may require the rotation of duties, so this subroutine would perform this operation. Every time the lead duty pump stops, the duties could be rotated, so that the next time a pump is called for, the alternate or next pump would operate; this would ensure a balanced use of the field devices.

RW_Pump_Control

This routine contains the actual start/stop logic associated with the pumps. The availability and duty assignments would be incorporated into the logic so that the lead duty pump is operated in accordance with the duty assignments.

In most systems, there is allowance for both Remote Manual and Remote Automatic modes of control. In the former, the operator can start and stop the pumps; in the latter, the automatic control logic for the pumps would determine when each pump is to be operated.

Travel_Screen_Control

Like the raw water pumps, the travelling screens need to be operated to clean them, typically at regular intervals. Perhaps once per hour the travelling screens should be operated for possibly 5–10 min, then stopped. While the screens are moving, a water spray would be ejected against the screens to clean the debris that had collected on them. The operation of both the screens and their associated sprayer systems would be controlled here.

Current_Data_Monitor

The monitoring of process values such as flowrate, water turbidity (degree of dirtiness) and pH would be stored in the memory of the PPC. While there would obviously be database points for each such value, additional buffered values may be needed for collecting samples over time.

Process_Outputs

Like the process inputs routine which transfers field signal values to virtual software points, the virtual software output values need to be transferred to the field output signals. Again, the separation of the internal software points and the hardware field signals would allow for testing and troubleshooting without affecting the real-world outputs.

Process_Alarms

Alarm conditions which arise in the program logic must be processed and reported to the host SOW system. Some discrete inputs, such as Motor High Temperature, would indicate an alarm condition by virtue of becoming 'True'. Certain analog signals, which have crossed over the high or low alarm setpoints, would generate alarm conditions. This routine would therefore perform the necessary processing or handling of the various alarm conditions that may arise.

9.3.4 Defining the Synchronous and Asynchronous Logic

As described previously, there would be a synchronous or periodic task and program to handle the operations which must be performed at fixed time intervals. One of the most common examples of this would be the execution of a PID control loop for closed loop control. In order for the integral function to work properly, the PID instruction must be executed repeatedly at fixed time intervals, such as 10 times per second.

In our sample application, three processes were identified which might best be handled using a synchronous task and program: flow totalizing, pump runtime accumulation and communication with other PPCs, such as the Pretreatment PPC mentioned earlier. A single synchronous task with one program could be defined which would include subroutines for each of the fixed interval processes to be handled. The subroutines would be called by the mainline routine of the program. An illustration of this structure is shown in Figure 9.5.

The mainline routine, 'Collect_Data', would call the various subroutines to perform the time-based operations indicated. A brief explanation of each of the subroutines is as follows.

Flow_Totals

By summing the current flowrate at regular intervals, a total flow volume can be calculated. For every interval, such as 10 times per second, the current flowrate would be divided by 10, then added to a total flow volume. This summed flow could be divided by 1000, resulting in a total flow in cubic metres. This summing and converting would be done for any process values that require such totals.

Runtime_Accumulation

For maintenance purposes, it is useful to monitor the runtime of major pieces of equipment, such as pumps. This runtime may be accumulated in Seconds, Minutes and Hours, or in Tenths of Hours and Hours.

PPC_Communication

Data being transferred between PPCs does not have to happen continuously; such transfers can be programmed to happen at intervals such as once per second and twice per second. By placing the data transfer logic in this cyclic task/program, the transfer operations can be triggered at these regular intervals, thus reducing the amount of 'traffic' on the PPC network.

The specific data transfer requirements would be programmed into this routine so that the data transfers happen at the fixed intervals needed for the application.

In addition to the cyclic or periodic type task, there is also an asynchronous or event task which is triggered upon some condition. Typically, there may be one or more

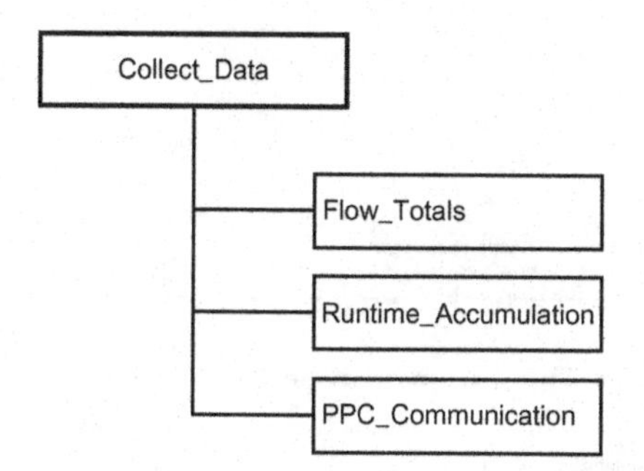

Figure 9.5 Synchronous/periodic program structure with process routines.

discrete input signals that require critical response when they change state; these would represent top priority alarm conditions. By configuring an event task to execute when one of the specified conditions occurs, program logic can be created inside this event task which responds to the particular requirements of the signals. In some cases, the occurrence of the input change may require the shutting down of equipment and/or issuing alarm signals.

9.3.5 Choosing the Programming Languages

One other consideration when programming PPCs is the choice of language. The traditional language used in PPCs in North America is Relay Ladder Logic, a graphical language using rungs or networks of code. This language is most efficient for boolean type operations and step-by-step controls. Today's PPCs typically offer a range of programming languages, in addition to the traditional Ladder Logic used in Canada and the United States. The IEC (International Electro technical Committee) established a programming standard for PPCs which includes five standard languages: Ladder Logic, Structured Text, Function Block, Sequential Function Chart and Instruction List.

Some manufacturers, such as Allen-Bradley, GE Fanuc and Siemens, offer this suite of languages. Depending on the manufacturer, individual routines within programs can be developed in different languages; thus, a 'number crunching' routine might be programmed in Structured Text, a sequential control routine could be developed in Sequential Function Chart and a PID control loop could be programmed in Function Block. The choice of language or languages to be used depends on the specific application and how best to implement the logic for the PPC.

Following is a brief explanation of each language; images of sample logic for each language are shown in Figures 9.6–9.10, with brief language description to each.

The *Ladder Logic* is graphical in nature and easy to follow; figure 9.6 illustrates this language. Program logic is constructed in 'rungs' or 'networks' of logic, with comments of text assigned to explain the logic. Sequential operations as found in

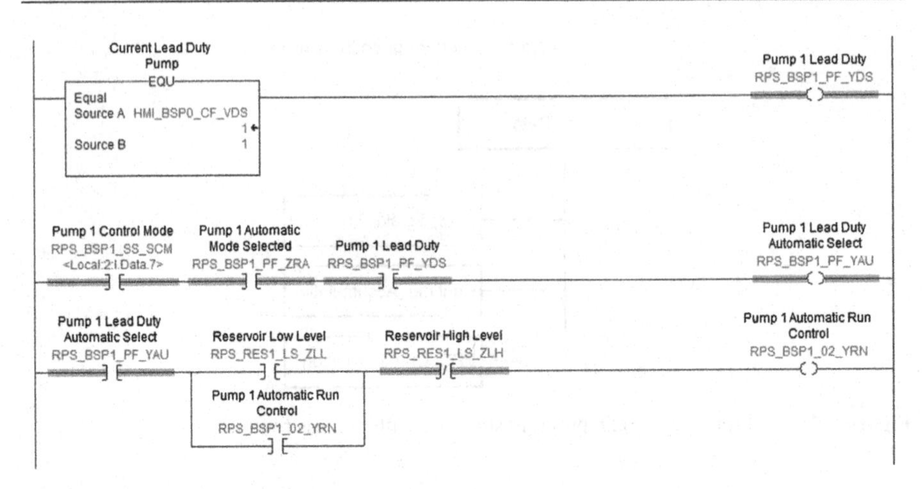

Figure 9.6 Ladder Logic programming language.

many systems lend themselves well to this language. This language is also easy to monitor as states of discrete points and values of analog points can be 'continuously observed' while the program is executing.

The *Structured Text* language is based on the popular 'C' programming language, which is a high-level general purpose language, as shown in figure 9.7. In applications which require considerable 'number crunching' and processing of data, this

```
IF Reservoir_Level  > 90.0 THEN

        Pump1_Run_Request := 0;  Pump2_Run_Request := 0;

END_IF;


IF Reservoir_Level  < 35.0 THEN

        Pump1_Duty =1  THEN

                Pump1_Run_Request := 1;  Pump2_Run_Request := 0;

        END_IF;

        Pump2_Duty = 1  THEN

                Pump2_Run_Request := 1;  Pump1_Run_Request := 0;

        END_IF;

END_IF;
```

Figure 9.7 Structured Text (high level) programming language.

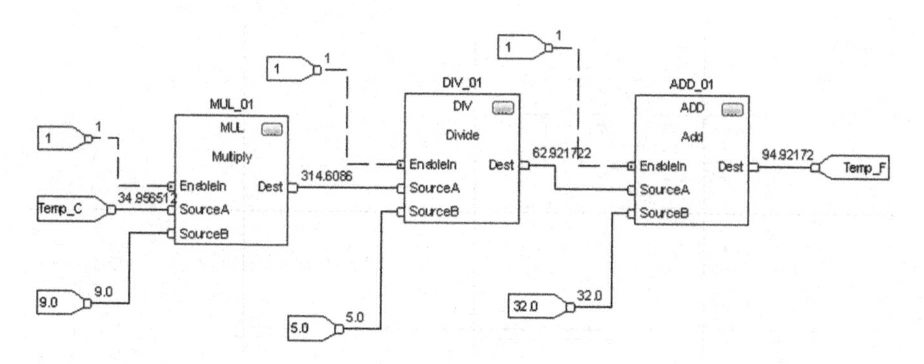

Figure 9.8 Function Block diagram programming language.

language is very efficient and easier to use than ladder logic. Operations such as calculating the surface area and/or trigonometric values simply need a single statement line, compared to ladder logic which might require several instructions. In the example below, the reservoir level is being compared with limits to determine which pump should be turned on or off.

The *Function Block* language is based on logical blocks of operations, interconnected with lines. This is a graphical language similar to Ladder Logic but is shown in free form without structured rungs or networks, as shown in figure 9.8. Each block is configured with inputs and outputs, and the programmer can create his or her own function blocks as desired. The current values of the inputs and outputs can be viewed while the program is executing. In the following example, the input temperature in degrees 'C' is being converted to temperature in degrees 'F'.

Sequential Function Chart would appeal to programmers who implement sequential step type logic; figure 9.9 illustrates this language. If a machine must step through a series of actions, with each step proceeding when the previous step has been completed successfully, then the sequential function chart is quite efficient. Each box in the flowchart defines one step, and the condition or conditions to move on to the next step are identified. A filter backwash sequence requires many sequential steps, so this language might be a suitable choice for this operation. The following example comes from one of the RSLogix 5000 illustrations.

Finally, *Instruction List* is a low-level machine language, in which each instruction has a unique mnemonic representing a single operation; figure 9.10 illustrates instruction list programming. Examples of instruction include add, multiply, load and store, branch on condition and call another routine. This language is used when maximum execution speed is desired, as the programmer can perform operations in very efficient ways that the other languages cannot offer. This language is often used when there is a special interface to the PPC which requires 'bit bashing' of bits within words of data. The following example sets and resets the output 'LP1' depending on inputs 'SW1', 'SW2' and 'SW3'.

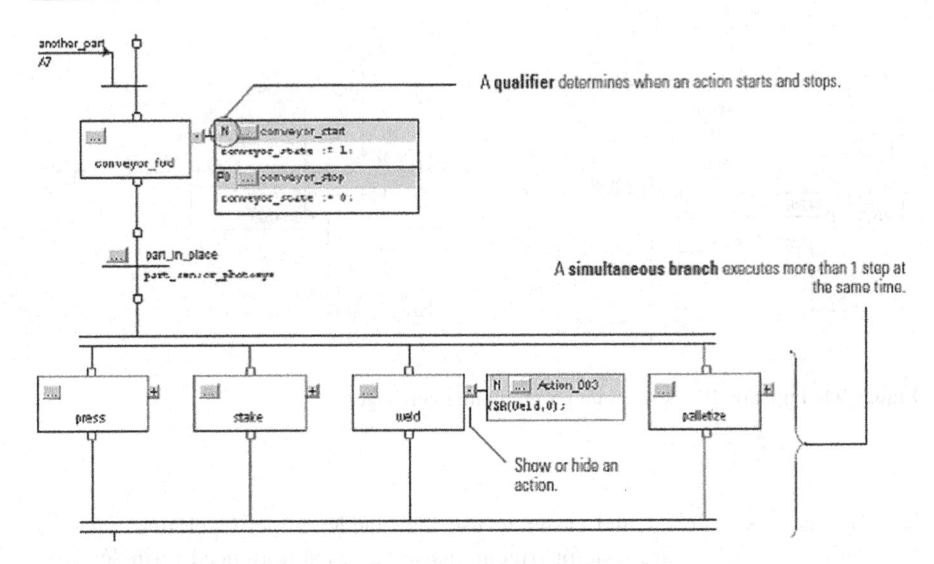

Figure 9.9 Sequential Function Chart programming language.

```
A     "SW1"                                                         I136.0
FP    "FlagBit"                                                     M0.0
S     "LP1"        // SET output with 'SW1' input                   Q136.0
O     "SW2"                                                         I136.1
O     "SW3"                                                         I136.2
R     "LP1"        // RESET output with 'SW2' or 'SW3' inputs       Q136.0
NOP   0
```

Figure 9.10 Instruction List or Machine Language programming language.

9.4 Programming Guidelines and Style

While the programming tools available today for controller programming are quite advanced, and contain many time saving features, the process of designing and developing program logic is still very much an art. No two programmers will program the same way. There are certainly some well-tested methods that most programmers use, but the actual program logic varies in how it is implemented.

It is the intent of this section to offer some suggestions on approaches to programming that save time and result in better structured and easier to understand program logic. After having worked in the industry designing and programming systems for more than 35 years, I believe I have some real-world experience to offer the reader.

The topics presented in sections 9.4.2 to 9.4.6 include: Diagnostic Features, Top-Down Structured Programming, Function Block Program Design, Database Maintenance Using Spreadsheets and the Iterative Design and Program Process.

9.4.1 Guidelines for PPC Programming

As much as programming software has been improved with new features to make program development easier, the actual programming process is still very much an art. No two programmers design and code program logic the same way; the important point is that in the end, the application program correctly performs all of the operations defined in the PCLDs. Some programmers prefer to use very few subroutines, possibly entering all of the logic in the one mainline routine; other programmers prefer to structure the application using many subroutines, sometimes with three or four levels of routines. This is an example of where the programming style and the art of programming enter the picture.

One important aspect to PPC programming is the use of descriptions and comments in the logic. While the use of structured tagnames as described in Chapter 5 results in more descriptive point names, additional documentation of the program is always helpful. Descriptions should be added to the tagnames for a more complete definition of the program point. In the case of A-B ControlLogix controllers, the tagname, hardware address and the descriptions all get downloaded to the processor, so going online to a controller shows all of this information.

Consider the illustration of some program logic for establishing the mode of control. Figure 9.11A shows logic with only the tagnames, in which the tagnames follow a structured approach. If one understands the tagging system, one can determine what the program logic is doing. Figure 9.11B shows the same logic with operand descriptions, and as one can see, the meaning of the software points is noticeably clearer.

Rung comments should be used to describe the collective logic in a section of the program. For example, the logic for starting and stopping motors may require 5–10 rungs of code; an explanatory comment should be inserted at the beginning of this code which explains the concepts and logic which follows.

Figure 9.12 illustrates program logic with a rung comment; this comment explains the purpose and function of the next two rungs of logic. Specifically, virtual points triggered from the HMI will set and reset the Manual and Automatic mode points, which are then used later in the pump control logic.

Creating subroutines for processing inputs, processing alarms and processing outputs allows the programmer to test the logic without affecting the real-world signals. The programmer can set and clear points in the logic to simulate the inputs and thus test how the rest of the logic operates. I have used this technique extensively, such that when I actually arrived on site, I have already verified much of my program logic by simulating points in the program. The same setting of values works for analog points, so this technique is not limited to discrete type points.

9.4.2 Programming of Diagnostics

Today's PPCs include a number of diagnostic features which can be incorporated into the program logic. The application program is designed to handle all of the

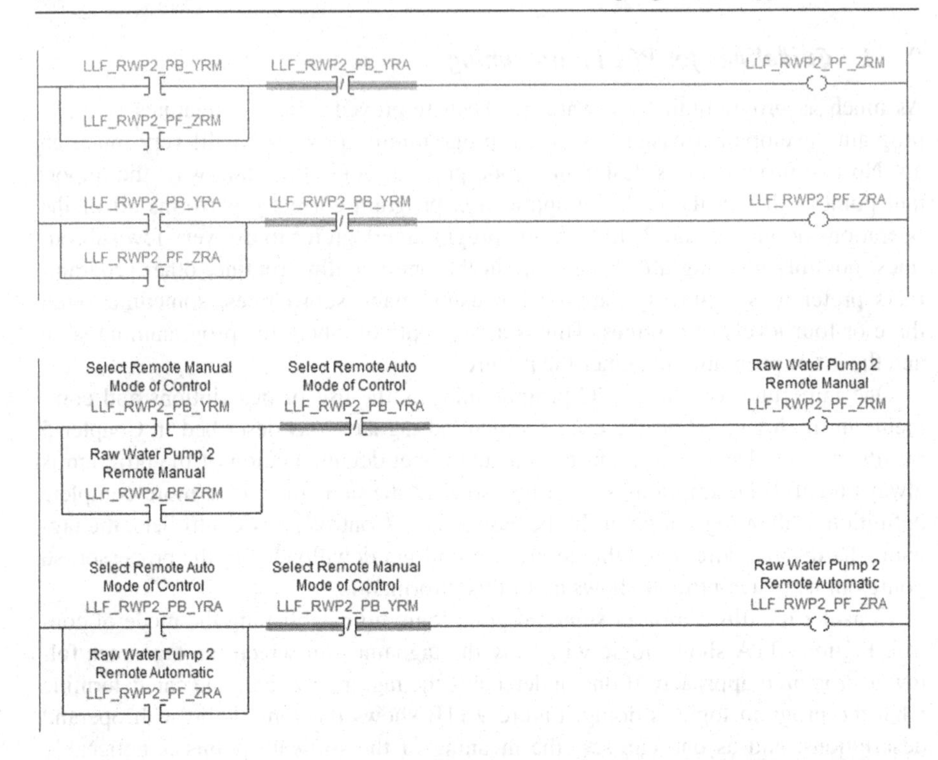

Figure 9.11 (A) Logic without operand descriptions. (B) Logic with operand descriptions.

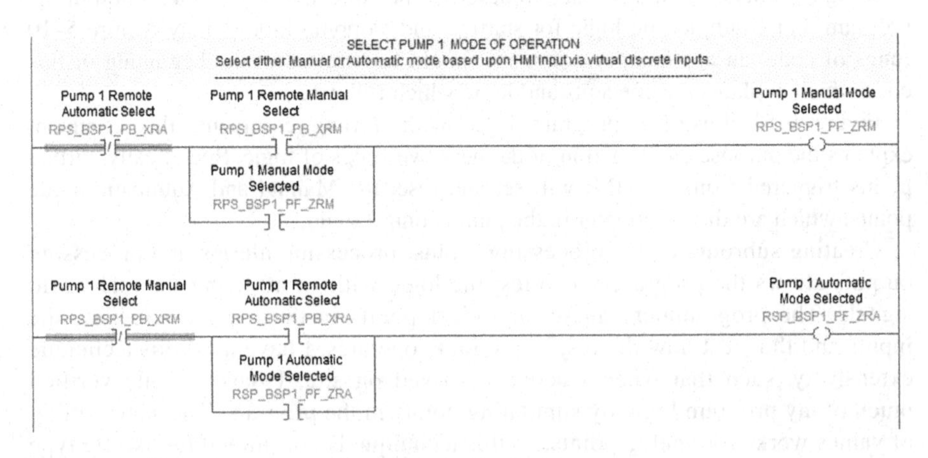

Figure 9.12 Manual and Automatic mode selection logic.

operations required of the PCLDs, including handling errors and abnormal conditions. After all, the PPC was designed to operate 24h a day, 7 days a week, without any intervention by the user. But how can the built-in diagnostic features of the PPC be used to report problems to the user via the SOW?

Using the A-B ControlLogix controllers as an example, the diagnostic discrete input and output modules include a number of configurable diagnostic features. The input module can report if any of the inputs has an open wire condition; that is, the continuity of the input signal has been lost, most likely due to a broken wire or other form of disconnection. For the discrete output modules, the module can report if any of the outputs has an open circuit, referred to as a 'no load' condition. The output status is compared to the status sent to it by the program logic; if the two statuses do not match, then a 'output verification failure' error condition is reported. For both the discrete input and output modules, a fault on any point can be reported.

The analog input signals, which can be scaled in engineering units within the module, can report when the raw input or output signal is out of range; for example, a 4–20mA signal is below 4mA or above 20mA. With alarm limits configured for the analog inputs, an alarm condition is reported when the scaled value crosses one of the alarm limits. Also, if the analog value changes too quickly, a Rate of Change error is reported.

All of these diagnostic conditions can be detected by the user program in such a way that the error conditions can be passed on to the SOW to alert the operator of the system. As an example, each of the 'open circuit' bits can be multiplexed into a single integer which is passed to the SOW; there, the bits can be demultiplexed into visibility type indications on a screen, as well as causing an alarm if any of the bits is non-zero. Similarly, conditions such as rate of change alarm and crossing alarm limits can be reported and displayed to the SOW operator.

9.4.3 Top-Down Structured Programming

The concept of Top-Down Structured Programming is not a new concept; this approach has been championed by many programmers in the past using a wide variety of programming languages. This approach is to develop the overall structure of a program with respect to the routines required, then to define the organization or calling structure for the program. Developing the program starts at the highest level of concept and works down into the details of each routine.

Starting with a mainline routine, a subroutine file must be created for each routine in the program design. A dummy or blank subroutine is created with only a 'Return' instruction inside. The mainline routine consists of 'Jump To Subroutine' or 'Call' instructions to each of the subroutines defined. An enable contact, normally open type, would precede each subroutine call so that the calls can be enabled one at a time. In this illustration, tagnames of the form 'Enable_SRx' are used to enable or disable the subroutine calls. A sample mainline routine is illustrated in Figure 9.13.

Each of the subroutines would consist of a coil at the start and a 'Return' instruction to define the end. An illustration of the 'Process_Inputs' subroutine is shown in Figure 9.14.

With each of the subroutines containing a 'Return' instruction only, this project could be downloaded into the PPC processor and tested for basic program flow. Since the Enable contacts are virtual or software points, they can be set True or False within the logic while online to the PPC. This establishes the basic structure and flow of the application program.

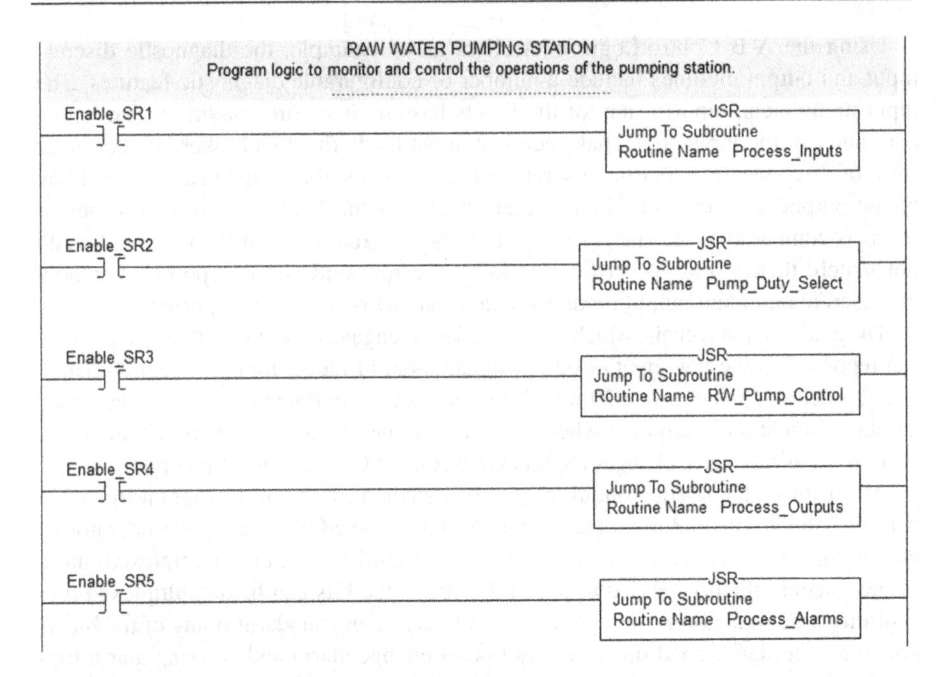

Figure 9.13 Initial mainline program logic.

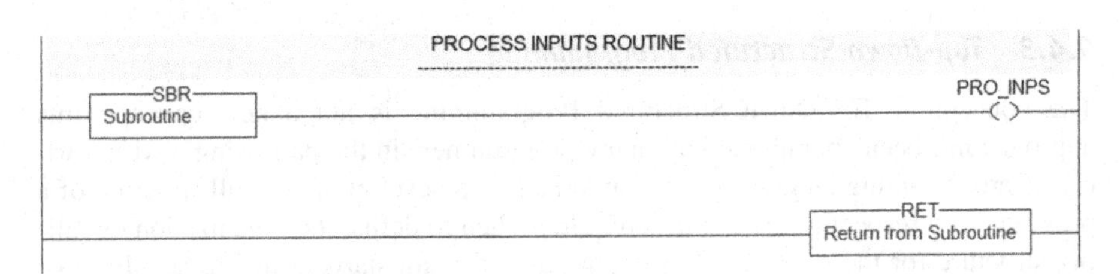

Figure 9.14 Initial subroutine logic66.

Referring to the PCLDs and the PPC points list prepared previously, the actual detailed program logic can now be developed for each routine in the program. As the program logic is developed, virtual software points will be required in the logic; these new points are simply added to the software points list for the PPC. For example, the Microstrainer will have a few field I/O points referring to the local/remote status, the running status and the run control signals. Additional software points will be added for the internal control mode, such as Remote Manual and Remote Automatic. Hence, the program logic and the database are developed together.

The software points should be assigned tagnames in accordance with the Tagname Signal Naming Convention (TSNC) established earlier in the project. As illustrated previously in this chapter, some software points for the Microstrainer were shown which followed the fragment structure defined for the project. By using this TSNC for all database points, regardless of the point's origin, the program maintains a consistent naming convention that follows an established standard. Later, one can read the program logic and recognize the tagnames by their fragments and understand how the program operates or functions.

As each subroutine's logic and database is developed, the program can be downloaded and tested, with all other routines being disabled. All such testing would be done without interacting with the real-world inputs and outputs. Once all of the routines have been coded and tested, a complete program will result which has followed a well-structured and documented design.

9.4.4 Program Design Using Function Blocks

One approach to developing the detailed program logic involves the use of function blocks for which the function is defined with the required inputs and the desired outputs. The PCLDs contain all of the details on what the application program must do; these documents, organized by PPC, describe the process operations in a conceptual way. The programmer must combine this information with the field I/O hardware points in order to develop the detailed program logic code.

One method of developing the detailed code which may prove helpful is to create function blocks with all of the inputs and outputs associated with the process being programmed. The function block is simply a rectangle with the input signals listed on the left side, and the output or result signals listed on the right side. This approach offers a graphical representation of the program logic. A point form summary of the logic within the rectangle can be written which describes the details of the program logic.

First, select one of the functions or processes to be programmed; this could be one of the subroutines identified earlier in Section 9.3.3 or 9.3.4. For illustration purposes, consider the programming of the 'Pump_Duty_Select' subroutine, which requires changing the duties of the pumps when a switch trigger signal occurs. A function block could be drawn with the required inputs on the left side, and the expected outputs on the right side. As an initial definition, consider the function block in Figure 9.15; the necessary input signals such as the Local/Remote statuses of the pumps are shown on the left, and the expected output signals are shown on the right. Note that the inputs and outputs can include both field I/O signals and software points.

The process shown indicates that the Local/Remote status must be in the Remote mode of control. The 'Change_Duty' signal is a virtual software point in the program which signals that the duties of the two pumps are to be swapped: the current Lead duty becomes the Lag duty, and the current Lag duty becomes the Lead duty. The output signals consist of a boolean and an integer for each pump, denoting the availability of each pump (boolean) and the duty number (lead = 1, lag = 2).

The availability flag requires logic to combine the Remote status input with any other enabling inputs, such as Emergency Stop and/or Alarm signals. For example, if the pump is in Remote mode and therefore True, the Estop is False, and any alarm signal is False, then the Pump Available flag would be True; if either of the two

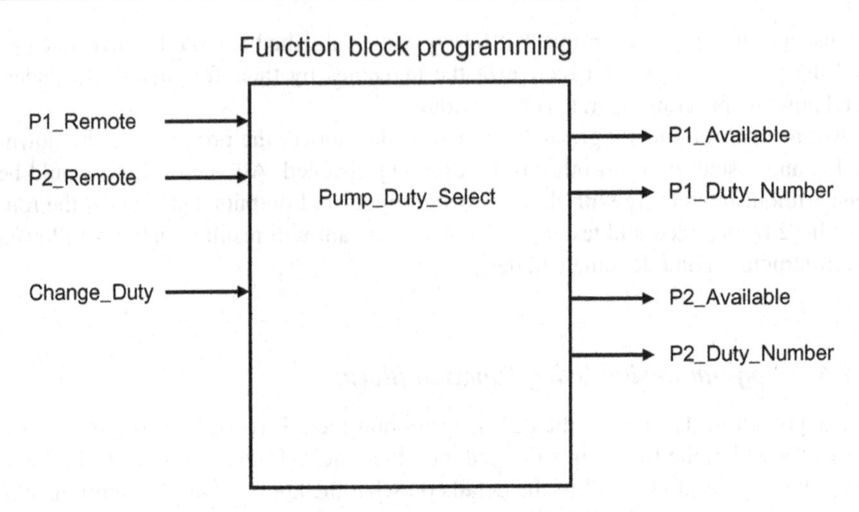

Figure 9.15 Function block for Pump_Duty_Select.

pumps is not available, then the duty switch cannot be made, and an error flag should be set.

Revising the function block to incorporate these changes, the resulting diagram would appear as shown in Figure 9.16.

Now all of the details required for the program module have been identified in the function block; this style of programming could be included along with the PCLDs and/or with the PPC Program Listing information, as a more detailed design document.

All of the necessary input signals have been listed so that the programmer knows what signals must be included in the program logic. All of the expected output signals have been listed so that the required results of the program logic are known. Note that some of the signals are boolean (flag) type and some signals are numeric (integer or double integer). These resulting signals can then be used elsewhere in the program logic, perhaps in other subroutines of the PPC program.

To illustrate a possible programming solution for this function block, consider the ladder logic shown in Figure 9.17. The logic is divided into three parts for clarification.

Figure 9.17A sets/clears the software flags required in the logic. There is a 'Pump Available' flag for each pump, and an enable flag if the pumps are available, and an error flag otherwise.

The second part performs the duty swap, provided the conditions are correct. Note how the duty numbers are switched for the two pumps, using a 'one shot' to ensure the change is only performed once when the 'Change Duty' signal is triggered elsewhere in the program.

Finally, the third part sets and clears the duty flags and the duty numbers for each of the two pumps.

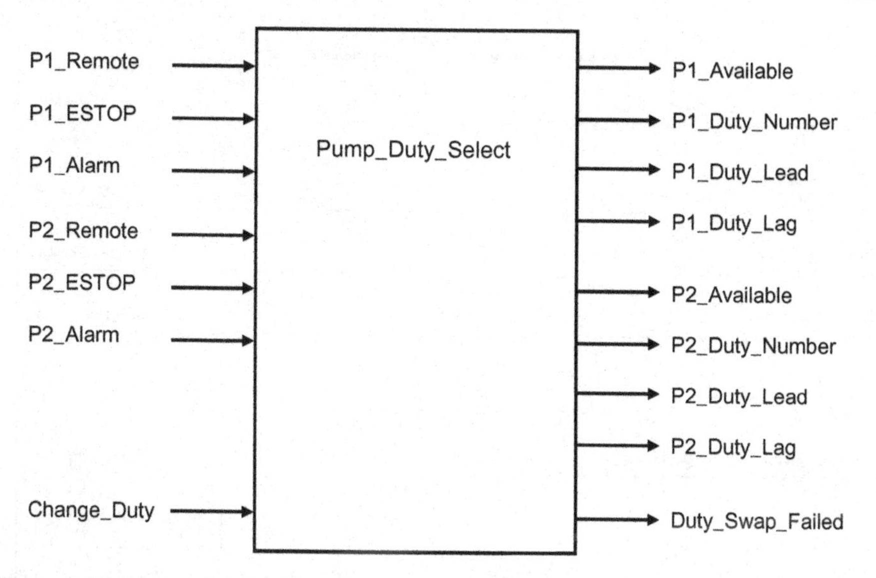

Figure 9.16 Revised function block for Pump_Duty_Select.

The preceding illustration shows the design of the function block and the program logic for the 'Pump_Duty_Select' subroutine of the 'Raw_Water_Station' PPC program. There are usually more than one way to implement the program logic, so this illustration simply serves to show one method of doing it. The important point in this section is to show how function blocks with inputs and outputs defined can be used for developing the detailed program logic for a PPC application.

9.4.5 Database Maintenance Using Spreadsheets

As explained in the previous section, the database points will be defined and created as the program logic is developed. It is important to maintain the database points lists in spreadsheet form so that all of the points in the program are included. Chapter 6 described a process of creating tags and exporting them to spreadsheets; from there, additional points could be added, and the resulting points lists could then be imported back into the application program. At any time, the spreadsheets can be reviewed and modified as necessary, then imported into the program.

The export/import form of the database is a CSV file, which can be imported directly into a spreadsheet file or workbook. The workbook can include headings and formatting to make the database more presentable and easier to read. The points can be rearranged in any way that makes the project easier to understand and follow.

Typically there would be spreadsheets for the field I/O hardware points and for the internal software points; these two groups can be included in a single workbook if desired to maintain all database information in a single file. Alternatively, one could maintain two files, one for hardware points and one for software points. All of these points can be exported to CSV files, which in turn can be imported into the

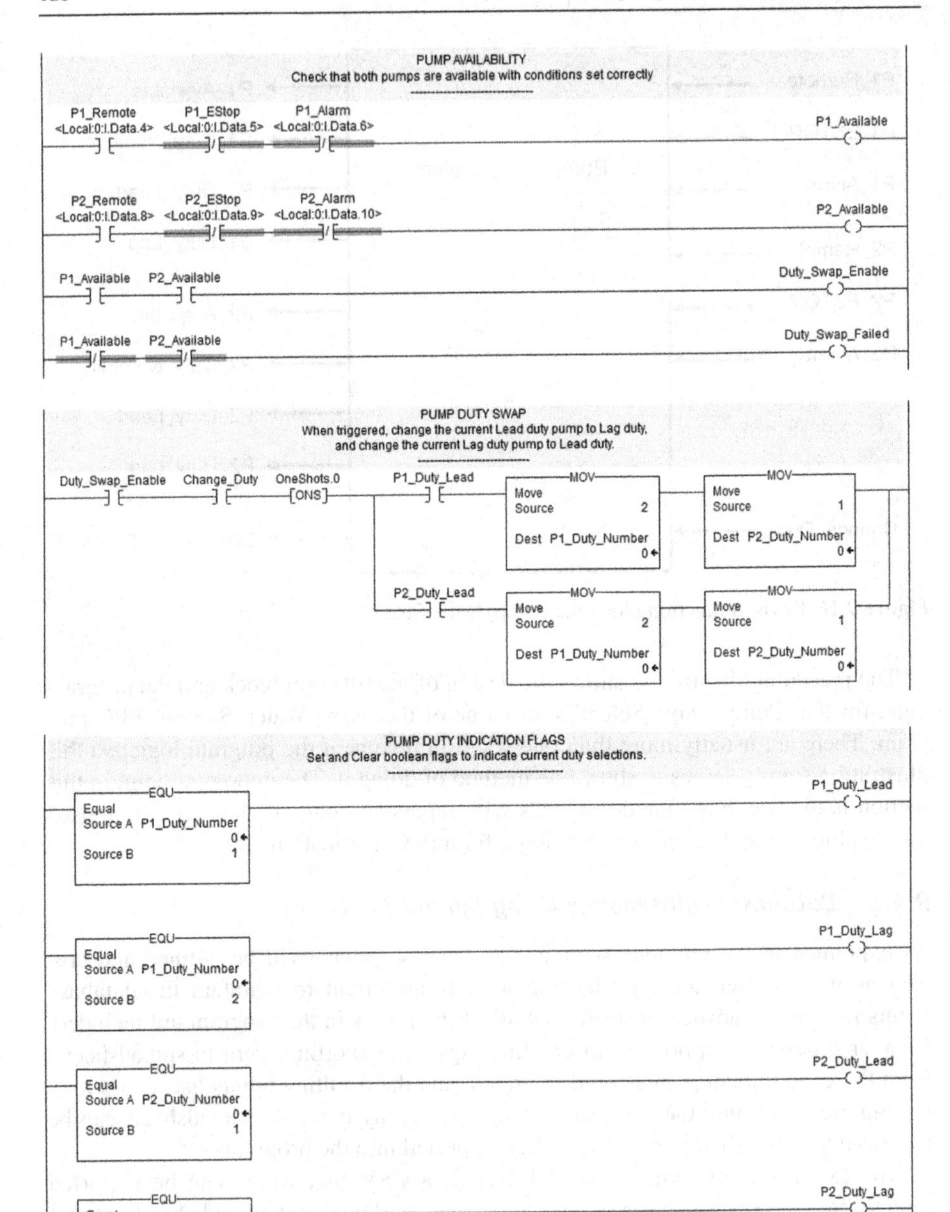

Figure 9.17 (A) Logic for pump availability. (B) Logic for switching the pump duties. (C) Logic for establishing boolean flags for the duty selections.

application program. Programming software packages allow for complete overwriting of the tag database, as well as updating with new entries.

9.4.6 An Iterative Approach to Programming

The process of developing program logic in subroutines, creating the database points and maintaining the documentation becomes an iterative process. This way, it is not necessary to define every point at the beginning of the project, since points will be needed as the logic is developed. The program logic and the database points are directly related, and one cannot have one without the other. So creating and documenting the database as the program is being developed ensures good documentation once the project is complete. But this does require an iterative process of creating logic, creating new tags, documenting the tags, creating more logic and so on. Just as the spreadsheets should be kept up to date, any changes to the logic design should also be noted in the PCLDs; quick notes would be sufficient during the programming, with the detailed editing being done later in the project.

On a few projects that I completed, I first created the tag database points in spreadsheets, then copied the content to CSV files and imported them into the application program that I was developing. This approach ensured that I was assigning tagnames in accordance with the defined tagging system, while at the same time, checking that I did not duplicate tagnames. As the programming process continued and new points were required, I added them to both the spreadsheets and to the program database. Periodically I would export the entire program database to a CSV file, then compare it against my documented spreadsheets to make sure they both agreed. Any differences were addressed and corrected, and the process continued.

Using this iterative method and maintaining the spreadsheets meant that when the program project was complete, I had accurate database spreadsheets that reflected exactly what is in the application program. These spreadsheets then became the final documentation for the project, along with the PPC program listings and the PCLDs.

9.5 Some Final Notes on Programming

The descriptions in this chapter are intended to provide a comprehensive and structured approach to designing and developing application software for PPCs in automation systems. While I have applied this methodology to all of my projects, I am not suggesting that everyone should follow my approach; rather, I am offering these methods as ways to create your own programs, while still using your own programming style.

I realize that programming is still as much an art as it is a science, and that everyone has their own methods and styles of programming. From my own experiences, I can say that the methods described herein have always resulted in well-structured and well-documented software that customers and consultants have both appreciated and commented on. In addition, I know that I have a detailed record of both the program logic and the controller database at every step of the software development.

One final suggestion on the approach to programming is to consider the 'what ifs' in the logic: for example, what if the pump/motor does not start? Since these controllers are designed to operate continuously without intervention, then we must make sure that our programs have considered every possible scenario. For every action that we program, we should consider the logic for the 'what if it doesn't work' case. Over the years I have found that the '80/20' rule still applies to any programming system: 80% of the time, 20% of the logic is being executed; and 20% of the time, the other 80% of the logic is being executed to handle the 'what if' scenarios.

In the example of a pump starting, what if the motor starts but there is no flow? In water systems, a flow switch is often used downstream of the pump to confirm that not only did the motor of the pump start, but that there is also actual flow through the pipe. Part of the check logic here would be to check that a set time after the motor has started, that the flow switch has been triggered. If a flow meter is used, then the flowrate should be above a minimum amount.

In summary, always consider the abnormal scenario, the case when the intended action does not happen. It is this extra logic that can prevent a system from malfunctioning as a result of some condition not being met, or some piece of equipment failed to do what is was programmed to do.

The methods outlined herein for software design and development do take a little more time, but the end results are certainly worth the end product. And as was pointed out in section 3.1 of Chapter 3, having the software documentation kept up to date provides ready material for submissions on contracts.

10 Guidelines for Workstation Application Programming

Objectives

- Identify the categories of information display
- Define the graphic displays conventions and navigation
- Define method of presenting and processing alarms and events
- Define the method of displaying trend information
- Design the organization of the historical database and its access method

The SCADA workstations incorporate both process graphic displays and databases; the software for these workstations requires the design and development of applications for different parts of the Human–Machine Interface (HMI). Also, the communication driver(s) for accessing the PPCs must be configured for optimum operations.

10.1 Identifying the Process Areas

From the previous chapter, the specific processes associated with each of the controllers was identified. For the workstations, the display and information requirements must be identified so that the necessary displays and functions can be designed. The purpose of the SCADA Operations Workstations (SOWs) is to present the current operating information about the system by retrieving information from the various Programmable Process Controllers (PPCs) in the SCADA system.

Typically, there is one PPC per process area. In a water treatment plant, for example, there might be four process areas, each of which would have a PPC programmed to implement the various processes in that area; the process areas could be

- Raw Water or Low Lift Station
- Pretreatment
- Filtration
- Treated Water or High Lift Station

One water treatment plant that I programmed included a packaged pretreatment system with its own integrated PPC; this PPC was programmed specifically to monitor and control all of the equipment within the package system. A second PPC was programmed by me to handle several other processes in that area of the plant, such as chemical injection systems, influent flowrate control, chlorination and effluent

Designing SCADA Application Software. DOI: http://dx.doi.org/10.1016/B978-0-12-417000-1.00010-5

flowrate control. There was a requirement to access some information from the packaged system PPC, so data was transferred between the PPCs using the messaging feature of the controllers. It is, however, more common to have only one PPC per process area of a plant or system.

10.1.1 Correlating Process Areas and Displays

Each PPC in a SCADA system is programmed to handle all of the input and output field signals related to the process operations in that area of the system. For the SOW, process graphic displays must be created which provide the operator with information about all of the processes related to that PPC. There may be more than one display for a given area, depending upon the number of individual processes; usually there would be an overview display plus additional detailed displays.

Returning to the water treatment plant application, the Raw Water or Low Lift Station PPC would monitor and control processes such as raw water screening, pump control, effluent flowrate control and the monitoring of various water quality data such as temperature, turbidity and well level. Process graphic displays would be needed to mimic the station operations, using colour changes, visibility of information and graphical animation. A general overview display would show all of the pumps, the inlet raw water well, effluent flowrate information and the key water quality parameters. An illustration of a draft overview display is shown in Figure 10.1; note that this could be a hand-drawn figure, as the actual creation of the displays is completed later, after all the required information has been identified.

This example shows the status of each of the pumps (running/stopped), the level in the raw water well, the effluent flowrate control valve position and the control modes of the pumps. The displays might have links incorporated such that clicking on a pump, for example, would display a pump control panel pop-up display; once done with this pop-up, an 'Exit' button would then remove the display from the screen. More will be said about these pop-up and other displays.

10.1.2 Identifying Information for Displays

The points lists produced earlier for both the field I/O hardware signals and the internal software points would be reviewed to identify what information associated with this PPC should be included in the displays. From these points lists, the programmer would identify which points should be included on the displays and how each should be shown.

For example, the raw water well level would be shown both as an analog value and as a graphic animation of the level in the well image; the pump run status would be shown with colour change to the pump images, making the image red for stopped and green for running (or reversed, depending upon preferences).

From these points lists, the programmer can prepare a list of points to be included in the displays for this process area. Some signals, such as alarm inputs, may appear in the Alarm Summary display rather than the Raw Water Station overview display.

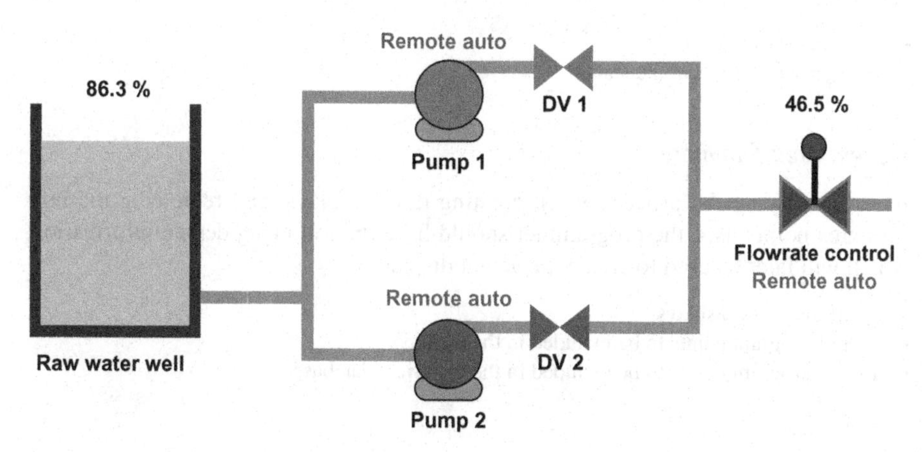

Figure 10.1 Sample draft overview display.

Referring to the draft display in Figure 10.1, the following list of points might be included:

- Raw water well level
- Pump operating status (for each pump)
- Pump operating mode – Manual, Automatic, Local
- Effluent flowrate
- Per cent position of the effluent valve
- Flowrate exiting the station and the flowrate setpoint.

In a later section, the details of how this information is displayed and how the display images can be animated will be covered. This procedure of documenting the required information must be completed for each process area. For some areas, there may actually be more than one display required in order to show all of the information on the SOWs.

10.1.3 Defining Historical Data Requirements

Historical trend displays are very useful for observing how data values change over time. Signals such as well levels and flowrates are typically trended. The process database contains the current values of every signal in the system, including both field I/O and software points. The historical database contains those signals for which trending is required; only a subset of the points needs to be trended.

One must review the points lists and select those points for which historical trending is needed. The selected points are then added to the historical database, with the frequency with which the data is to be sampled. For example, some values might be saved every second, while others are saved every minute. The raw water system illustrated previously might have the following signals added to the historical database:

- Raw water well level
- Effluent flowrate from the station.

Process Area Summary

After identifying the process areas, creating draft displays, and reviewing the process area points lists, the programmer should have the following design information, which will later be used to create the actual displays:

- Draft overview displays
- List of program points to be included in the displays
- List of program points to be included in the historical database.

10.2 Configuring the HMI Application

The HMI represents the software executing on the workstations of a SCADA system. This HMI software generally has two major components: the I/O server and database manager, and the user interface with displays, trends and other human interface information. Software packages such as Wonderware InTouch, GE iFIX and Inductive Automation Ignition are all structured in this manner, with at least two major components.

Both parts can execute on a single computer, but it is preferable to have one workstation performing the server functions, and one or more workstations providing the user interface. The operator workstations are referred to as 'clients', as they interact with the server computer to obtain the information to be displayed and to issue commands through the server to the PPCs in the system.

Figure 10.2 illustrates a simple network with server and operator workstations, as well as some field controllers. This figure will be used as an example in the following descriptions.

10.2.1 Functionality of Server Workstations

The server, or sometimes the 'host' computer or workstation, operates in the background, in the sense that its functions are not seen by the users. The server software is responsible for communicating with the various PPCs in the system, as shown in Figure 10.2, and maintaining the process (current) and historical databases. All data between the field controllers and the operator workstations passes through the server computer, using the current process database and the historical database.

Note that the red arrows represent data being retrieved from the PPCs to the server, and the green arrows represent data being passed from the server to the client operator workstations.

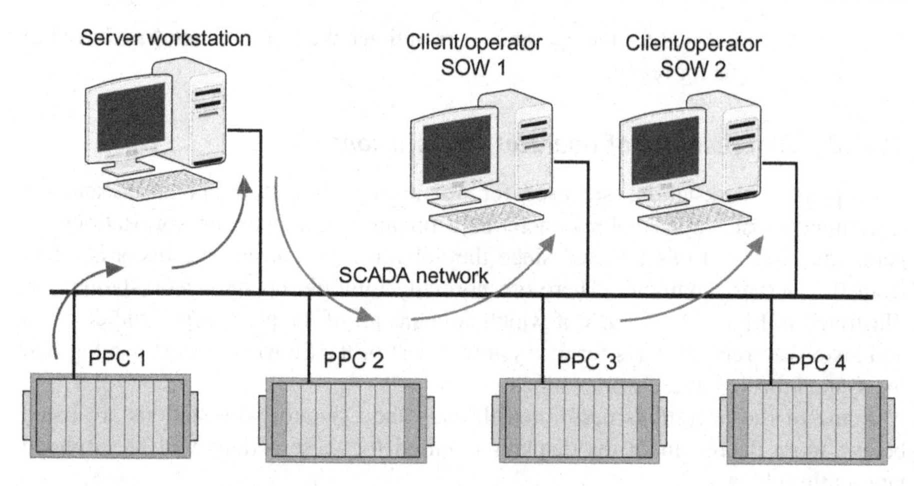

Figure 10.2 Illustration of SCADA workstations data flow.

The server workstation/computer performs a number of functions, which include the following major functions:

- Management of process and historical databases
- Communication with all PPCs to collect data and issue user commands
- Develop and modify application software for all workstations
- Interface with the WAN for transferring data files, often via the Internet.

The server computer/workstation maintains both a current or process database and a historical database. The process database contains all of the information from all of the PPCs in the system. The communication driver, or I/O server software, polls each of the PPCs to retrieve selected data values and updates this information in the process database. Any commands from the operator workstations, such as set-point changes, control mode selections or start/stop and open/close commands, are sent out to the intended PPC for effecting control.

Some data points, such as levels, flowrates and other process data, are also maintained in the separate historical database. This database records values for the specified data points at regular intervals so that trend displays can be generated using data that has been recorded over time. Each time the server stores a value for a point, the date and time are stored with it. The storage method, however, results in a compressed or compacted database to save file space.

Historical database files are usually organized as a Structured Query Language (SQL) database, in which SQL requests can be made to retrieve specific data values from the database. Information in this historical database is often sent on to other users external to the SCADA system. The SQL nature of the database makes it easier to share data with other applications executing on other workstations in the Wide Area Networks (WAN).

The server computer is typically the development computer for creating and modifying the application software for the specific SCADA system. All of the source files for the system are maintained on the server, so backup procedures should be in place to protect the software development. Once application software has been created and/or modified for the operator workstations, the files are then transferred to those workstations for regular use.

10.2.2 Functionality of Operator Workstations

The operator workstations are used for monitoring all system operations and for effecting control actions and parameter adjustments. These operator workstations are generally referred to as 'clients' since they obtain their current and historical data from the 'server' computer. There are normally multiple operator workstations, as illustrated in Figure 10.2, each of which contains all of the process graphic displays and historical trend displays for the system. Users of the SCADA system can log into the system through these workstations.

Some of the operations performed through the Operator workstations are listed below; more details about the displays required for these actions will be provided later in this chapter:

- Logging on and off the system using passwords and user names
- Invoking process displays to view the operations throughout the system
- Effecting control modes for various equipment in the system; for example, Manual and Automatic modes, placing equipment in or out of service
- Changing setpoint parameters, with appropriate security allowance
- Effecting manual control actions for equipment, such as start/stop and open/close
- Viewing historical trend displays and transferring data to other files for exporting
- Viewing the current alarm summary to identify alarm conditions requiring attention
- Viewing the alarm/event summary to view the chronological series of events.

The operator workstations provide the user interface or HMI to the SCADA system. Users can effect control over the equipment, as well as invoke displays which show current and historical information about any aspect of the SCADA system.

10.3 Developing the Process and Historical Databases

Before the process graphic displays and the historical trend displays can be created and animated, the process and historical databases must be created. Fortunately, the database can be created from the same spreadsheets that were used to create the controller databases. In Chapter 6 it was explained that the workstation process database is the combined databases from all of the field controllers (PPCs). Since points lists were created for the PPCs, and included both field I/O points and virtual software points, then these spreadsheets can be used to create one complete database, which includes all of the points for the entire system.

10.3.1 Creating the HMI Workstation Database

The HMI software package, whether it is the InTouch system, iFIX system, Ignition system or some other software, requires the data points to be in a different format from that of the controllers. But as was done to create the databases for the PPCs, the

:IOAccess	Application	Topic
RPS	DASABCIP	RPS-PPC
Galaxy	\\NA\NA	NA
CLX2	DASABCIP	Pumpstation

:IODisc	Group	Comment	OffMsg	OnMsg	AccessName	ItemName
RPS_BSP1_DV_SOD	$System	RPS Pump 1 DV Open Status	NotOpen	Open	RPS	RPS_BSP1_DV_SOD
RPS_BSP1_DV_SCD	$System	RPS Pump 1 DV Close Status	NotClosed	Closed	RPS	RPS_BSP1-DV-SCD
RPS_BSP1_00_DRN	$System	RPS Pump 1 Run Control	Stop	Run	RPS	RPS_BSP1_00_DRN
RPS_BSP1_SS_SCM	$System	RPS Pump 1 Control Mode	Stopped	Running	RPS	RPS_BSP1_SS_SCM

Figure 10.3 Sample export of HMI database points.

workstation database can be created by first entering a few sample points in the HMI software system, then exporting them to a Comma Separated Value (CSV) file. The rest of the points for all controllers can then be copied and pasted into the CSV file, since this file has the format required by the software package. The completed CSV file can then be imported back into the HMI software to create the database.

Referring to Section 6.4.1, some of the figures are reproduced here to illustrate the technique using spreadsheets. After some sample points have been entered into the HMI software and exported, a CSV file similar to that in Figure 10.3 will result. Note that this has the required format for the HMI software.

The remaining points can then be added to this CSV file; then import the file back into the HMI software. Many of the fields from the initial entries can be copies for the added points. This approach saves having to enter each point into the database using the HMI data entry facility and also ensures that all of the points in the system are included.

A spreadsheet of the CSV file after many of the points have been added is shown in Figure 10.4. Note that this illustration is a subset of the complete database and is being shown for illustrative purposes.

Rather than duplicating the methods and techniques described in Chapter 6 here, the reader is referred to that chapter for the detailed procedures and methodology for creating the process database.

:iODisc	Group	Comment	OffMsg	OnMsg	AccessName	ItemUseTa;	ItemName
RPS_BSP1_00_SRN	$System	RS Pump1 Run Status	Stopped	Running	RPS	Yes	RPS_BSP1_00_SRN
RPS_BSP1_SS_SCM	$System	RS Pump1 Control Mode	Local	Remote	RPS	Yes	RPS_BPS1_SS_SCM
RSP_BSP1_ 00_AGF	$System	RS Pump1 General Fault	Normal	Alarm	RPS	Yes	RSP_BPS1_00_ AGF
RSP_BSP1_DV_SCD	$System	RS Pump 1 DV Closed Status	Normal	Closed	RPS	Yes	RSP_BSP1_DV_SCD
RSP_BSP1_DV_SOD	$System	RS Pump 1 DV Open Status	NotCls	Open	RPS	Yes	RSP_BSP1_DV_SOD
RPS_BSP2_ 00_SRN	$System	RS Pump 2 Run Status	NotOpn	Running	RPS	Yes	RPS_BSP2_00_SRN
RSP_BPS2_SS_SCM	$System	RS Pump 2 Control Mode	Stopped	Remote	RPS	Yes	RSP_BSP2_SS_SCM
RSP_BPS2_00_AGF	$System	RS Pump 2 General Fault	Local	Alarm	RPS	Yes	RSP_BSP2_00_AGF
RSP_BSP2_ DV_SCD	$System	RS Pump 2 DV Closed Status	Normal	Closed	RPS	Yes	RSP_BSP2_ DV_SCD
RSP_BSP2_DV_SOD	$System	RS Pump 2 DV Open Status	NotCls	Closed	RPS	Yes	RSP_BSP2_DV_SOD
RPS_RES1_LS_ALH	$System	RS Reservoir High Level Alarm	NotOpn	Open	RPS	Yes	RPS_RES1_LS_ALH
RPS_RES1_LS_ALH	$System	RS Reservoir Low Level Alarm	Normal	Alarm	RPS	Yes	RPS_RES1_LS_ALL
RSP_BSP1_00_DRN	$System	RS Pump 1 Run Control	Normal	Alarm	RPS	Yes	RSP_BSP1_00_DRN
RSP_BSP1_DV_DON	$System	RS Pump 1 DV Open Control	Stop	Run	RPS	Yes	RSP_BSP1_DV_DON
RSP_BSP1_DV_DCE	$System	RS Pump 1 DV Close Control	NotOpn	Open	RPS	Yes	RSP_BSP1_DV_DCE
RSP_BSP2_00_DRN	$System	RS Pump 2 Run Control	NotCls	Close	RPS	Yes	RSP_BSP2_00_DRN
RSP_BSP2_ DV_DON	$System	RS Pump 2 DV Open Control	Stop	Run	RPS	Yes	RSP_BSP2_ DV_DON
RSP_BSP2_ DV_D CE	$System	RS Pump 2 DV Close Control	NotOpn	Open	RPS	Yes	RSP_BSP2_ DV_DCE
			NotCls	Close			

Figure 10.4 Sample of populated CSV file.

10.3.2 Creating the Historical Database

Historical databases are more often of the SQL form, which provides a more general data entry and retrieval facililty. Relational databases, such as used by companies to maintain information about customers and purchases, allow for access to information based upon different criteria. For example, a query might be requested for all points in the Low Lift area of the SCADA system between 12 and 13 of June 2012. Another query might be for the values of the Low Lift Well Level between midnight 12 and 13 of August 2010. So a SQL database must be created using the data entry tools of that database.

Most HMI packages now use this SQL style of database, since it is simpler to access information and easier to retrieve data for selected points in selected time periods. Also, the SQL-based database has become a standard in software systems. Remember that the historical database is used to store data values with time and date stamps so that the information can be retrieved by tagname, process area, alarm group and so on.

Using the points lists for the various PPCs, one might select or identify which points are to be historically accumulated. Storing all points would result in a very large file on disk, which in turn would make accessing the data slower. The historical database should only contain points for which historical information is required.

10.4 Design and Development of Process Graphic Displays

Animated colour graphic displays are used to present current operating information to the user through the user workstations. The general layout of the displays, colour convention used and navigation methods must be decided upon. Before designing any displays, some standards and conventions need to be established; for example, how the user will navigate from one screen to another, and what colours will be used to represent equipment status. These topics are covered in the following sections.

10.4.1 Establishing Standards for Displays

Depending upon the system, there could be anywhere from half a dozen displays to hundreds of displays which can be accessed from the SOWs. One needs to establish an intuitive method of navigating through these displays. Due to the variety of displays, and the types of displays, one needs to establish standards on appearance as well as navigation methods. The following sections address the establishment of standards for display navigation, display format, colour conventions and alarm notification methods.

Display Navigation

The displays on the SOWs consist of many different types: process graphics, trends, summaries, menus and so on. In order to make it easier for the user to navigate from one display to another, there needs to be some intuitive way of accessing information from the top level overview down to the individual detailed displays. Methods for navigating typically consist of one or more of the following techniques:

- Menu display – list of displays with buttons to click for access
- Quick Links panel – set of buttons on every display to access common displays
- Hyperlinks within displays to access other displays
- Hierarchy of displays for 'drilling down' to desired level.

The *Menu* display, as illustrated in Figure 10.5, consists of a list of displays organized into logical groups; the user would simply click on the display text to access that display. There may also be hyperlinks within displays that allow the user to reach more detailed displays, but the basic menu type display makes it simple to get to any display from this main menu.

Some systems use a 'Quick Link' panel, which includes pushbuttons to reach specific displays; this is illustrated in Figure 10.6. This method includes buttons to access the most frequently used displays. The alarm summary, for example, is one which users would want to access often, and from any display in the system. Likewise, trends and reports can be readily accessed from anywhere in the system. Typically this Quick Link panel would appear on every graphic display.

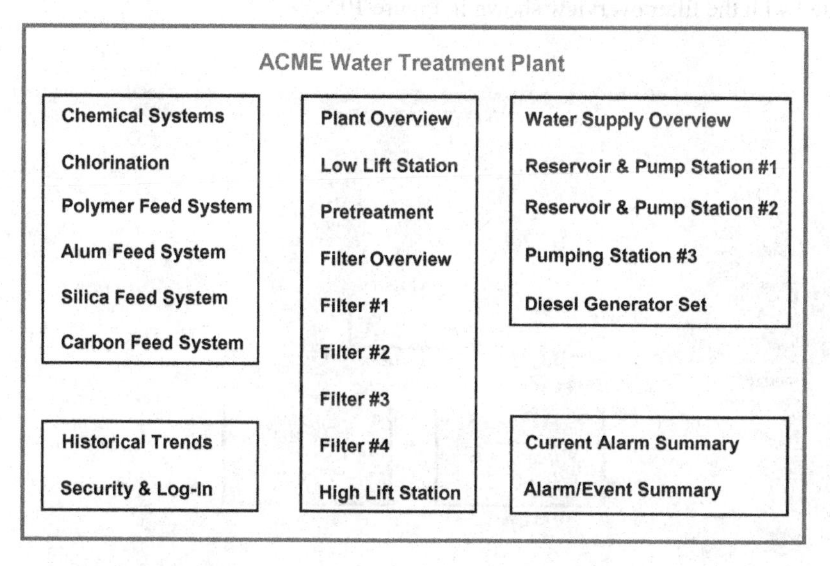

Figure 10.5 Sample menu display.

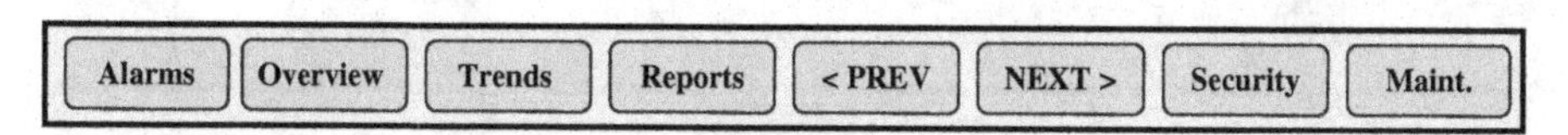

Figure 10.6 Quick links pushbutton panel.

Following is a brief explanation of each of the buttons illustrated:

- Alarms Alarm Summary and Current Alarm Summary
- Overview System or Plant Overview
- Trends Access to menu of all trend displays configured
- Reports Access to menu or list of formatted reports
- < PREV Previous display viewed in series
- NEXT > Next display to be viewed in series
- Security Log in and out; password setup
- Maint. Access to maintenance displays.

Hyperlinks within displays can be used to 'drill down' into more detailed displays; this is illustrated in Figure 10.7 for a water treatment plant overview display. Typically there is an overall display that contains some key information from the PPCs, such as levels in reservoirs and tanks, or status of conveyor lines.

There may be two or three levels of 'drilling' before one can get to the detailed screen desired. The links and buttons to accomplish this should be simple and straightforward. Consider the sequence of displays to go from the overview display to the display for changing the flowrate setpoint for a filter.

The overview display shown in Figure 10.7 shows the key information in part of the water treatment plant. By clicking anywhere on the filtration area, the user is presented with the filter overview shown in Figure 10.8.

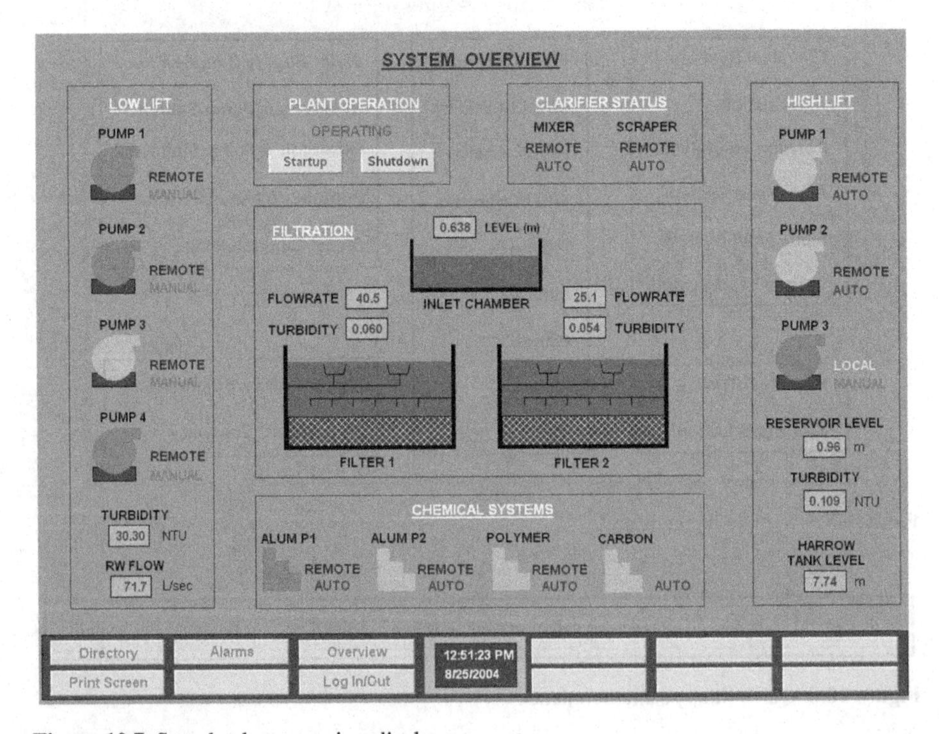

Figure 10.7 Sample plant overview display.

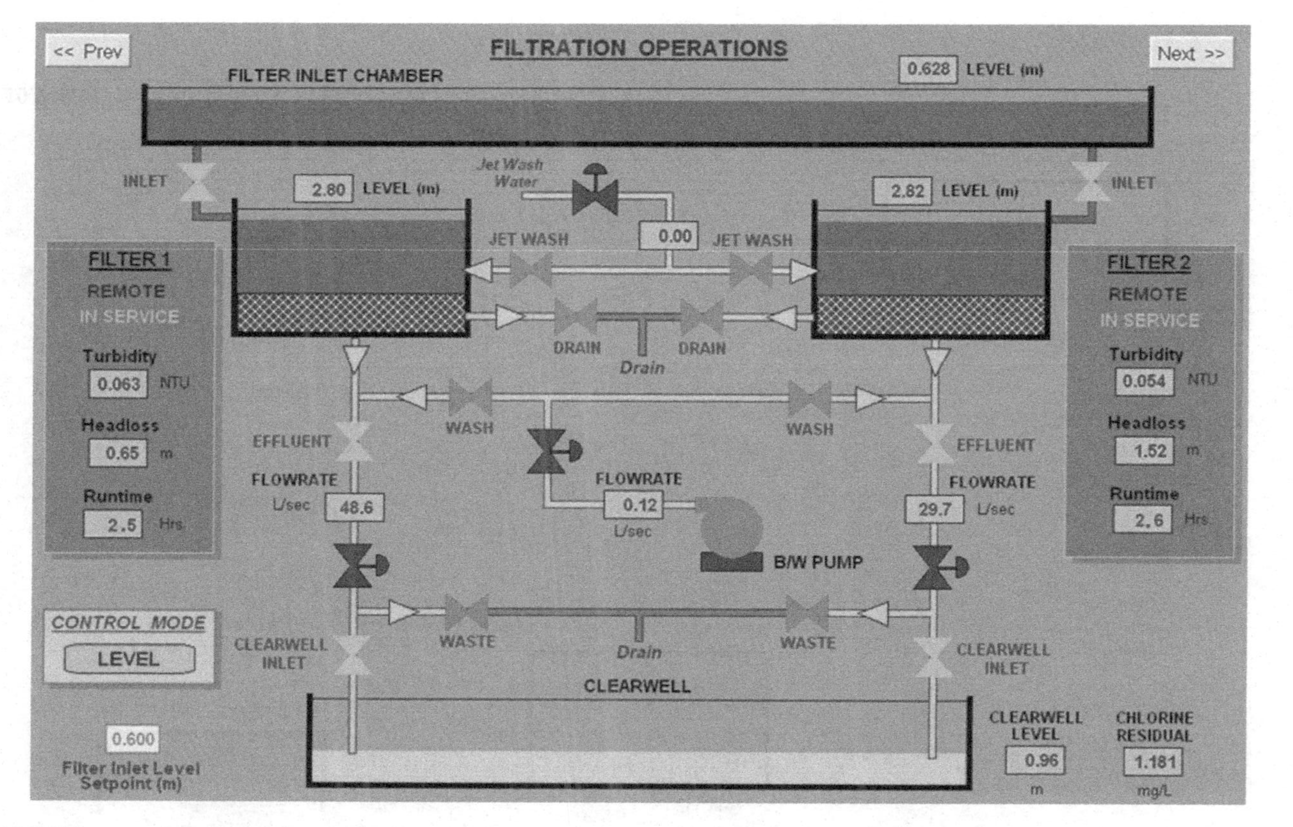

Figure 10.8 Filter overview display accessed from Plant Overview.

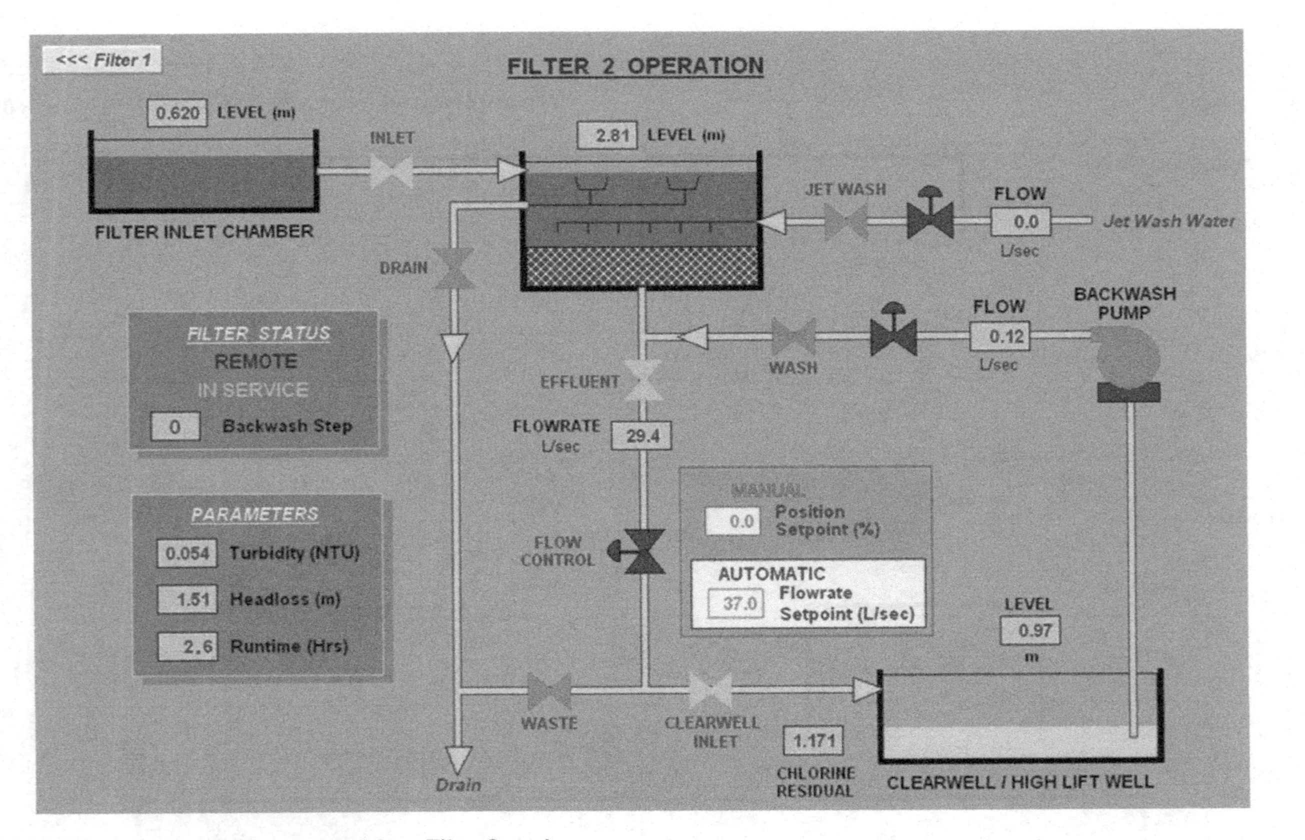

Figure 10.9 Filter 2 detailed display accessed from Filter Overview.

This filter overview display shows the current status of the two filters in the plant. All of the operations data for both filters is shown. In order to access the details for one filter, the user must click on one of the two filters, and the display shown in Figure 10.9 will appear; this is the detailed display for one filter.

Finally, by clicking on the 'Manual/Automatic' portion of the display, the mode of control can be changed, and the flowrate setpoint can be changed. This then represents one method of navigating through levels of displays, by creating display links in portions of the displays so that the user can work his/her way down through the levels to reach the specific information desired.

The structure or *hierarchy* of the displays should be considered, as there could be several 'levels' of display information. Consider a partial structure of displays for a water treatment plant as shown in Figure 10.10:

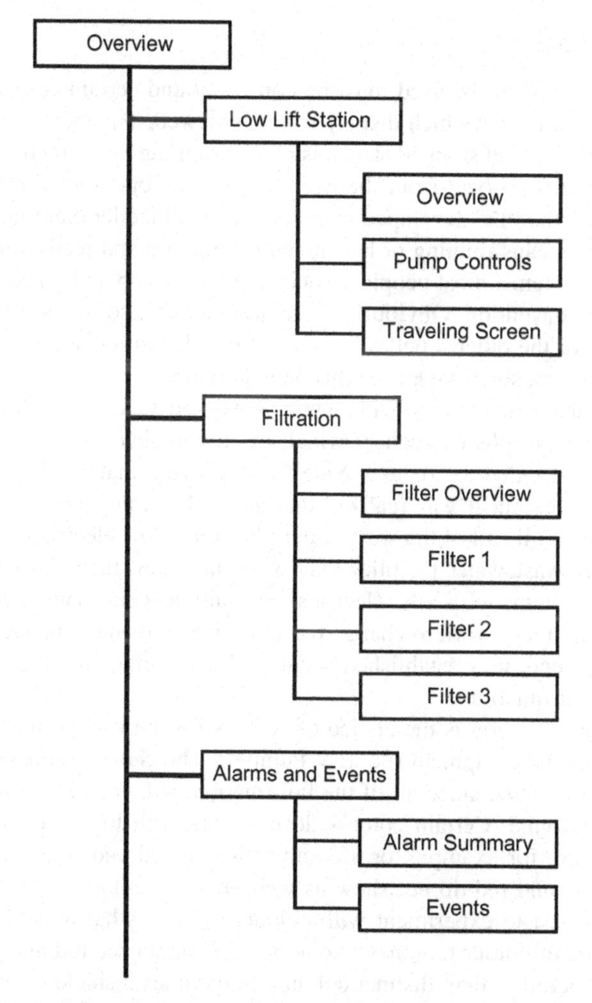

Figure 10.10 *Display hierarchy* – paths to move from one display to another.

In some systems, like the overview shown in Figure 10.7, there may sufficient displays that one overall menu is needed. Sometimes the main menu leads to submenus for each of the process areas. For example, selecting the High Lift pushbutton in the main menu might result in another menu showing all of the available displays associated with the High Lift portion of the system. The Pump Control pushbutton then may show a detailed display for the pumps and complete with the start/stop and control mode functions shown. I have worked with systems that had this approach, as the overall system involved many subsystems, each of which had many more displays.

Whatever methods are used to navigate the displays in a system, the method must be consistent and intuitive; the user should be able to get to the desired information without consulting a manual. Whether pushbuttons are used or hyperlinks in displays are used, the method must be consistent.

Colour Conventions

The choice of colours to be used must be consistent and certain colours must mean the same thing, no matter which display is being viewed. The most basic convention is that of the running and stopped states; is red for running or is green for running?

Process graphics evolved from the need to view the operations in the PPCs; the programming for the PPC developed from the electrical ladder diagrams. In the electrical world, red means running or hot; green means safe and ready for use. Outside of the SCADA systems, most people associate red with stop and green with go, as in the traffic light convention. Obviously these are opposite, so one must make a decision to use one or the other. I believe that most people are tending towards the 'traffic light' convention, since we all see this on a daily basis.

There was one municipality with whom I worked that has multiple water and wastewater treatment plants, each of which has a complete SCADA system. When the municipality decided to create a Wide Area Network that would allow access to any plant in the region, it was realized that all of the water plants used the traffic light convention, while all of the wastewater plants used the electrical convention. As there were more wastewater facilities than water facilities, they chose to change the water SCADA systems to use the electrical convention. This change, while certainly extensive, required less work to change the water plants than to change the wastewater plants. In the end, they established a standard that applied to all systems, regardless of location or function.

The next consideration is the choice of colours for the background, titles, equipment and text on the equipment (i.e. RW Pump 4). The choice of background colour affects all other colours, since all of the information being displayed will be against that same background. Certain colours do not work well together; green text on a black background, for example, or a combination of red and blue on a dark background. Magenta and red do not show as well on some colours; readability is most important. It is best to experiment with colours to 'see' what looks best. And then there is the issue of colour blindness: some people cannot see red and green, as they appear as shades rather than distinct colours. Sometimes a shade of grey is best for the background colour on displays; this is neither black nor white, but on off-white that is relatively neutral and works well with most other colours.

With the background colour and the running/stopped colours chosen, the next consideration might be that of alarms and variations of alarms. In most SCADA systems, a point can be in one of three states: new unacknowledged alarm, acknowledged alarm and return to normal. One combination might be

- *New unacknowledged alarm* Flashing yellow
- *Acknowledged alarm* Solid gold
- *Return to normal alarm* Solid cyan or solid magenta
- *Event (non-alarm)* Solid blue or solid yellow

New alarms should capture the user's attention, so the use of flashing colours could prove quite useful. Once the alarm has been acknowledged, the colour would change to a solid or non-flashing colour, but would still be noticeable. The return to normal condition, which means the alarm condition no longer exists, should be a neutral colour.

From my experience, I have found it best to experiment with colours to see how things look, and determine whether or not the information can be seen clearly. With the right combination of colours, one can see at a glance, the status of all equipment displayed standing some distance away from the screen. Once the colour standards have been selected, then everyone knows what colours indicate before approaching the display to see the details.

Display Format

All of the displays in a system should have the same basic look and feel, just as the word processor, spreadsheet and presentation software have in an office suite. In addition to the information to be shown in a display, it may be helpful to show the last few alarms, the date and time, and the name of the person currently logged into the system. This information can be arranged in a variety of ways, but consider the following styles. Following are a couple of examples of formatting the displays to include the key information.

Figure 10.11A shows the title bar with the current user name, the title of the display and the date and time; the body of the display then contains the details of the information being presented. In Figure 10.11B, the information is moved to the bottom but includes the last three alarms in the system.

The second example includes the last few alarms, and this can be changed to include more than three; however, too many alarms here would use more space, and therefore reduce that amount of space available for other information. It is important to decide on what should appear on all displays.

Note also that the 'Quick Links' panel could be included in each of the displays; this panel could be placed at the top or the bottom of the display page. One possible option, given today's computer screens having the 16:9 ratio, is to place the panel vertically, such that the buttons are one above the other, and off to one side. Since the width of the screen is wider than the older technology which gave us the 4:3 ratio, the extra space on the sides could be used for this panel.

Alarm Notification

The colour choices for alarm conditions have been considered previously, but one must still consider how new alarms and/or changes in alarm conditions should be annunciated.

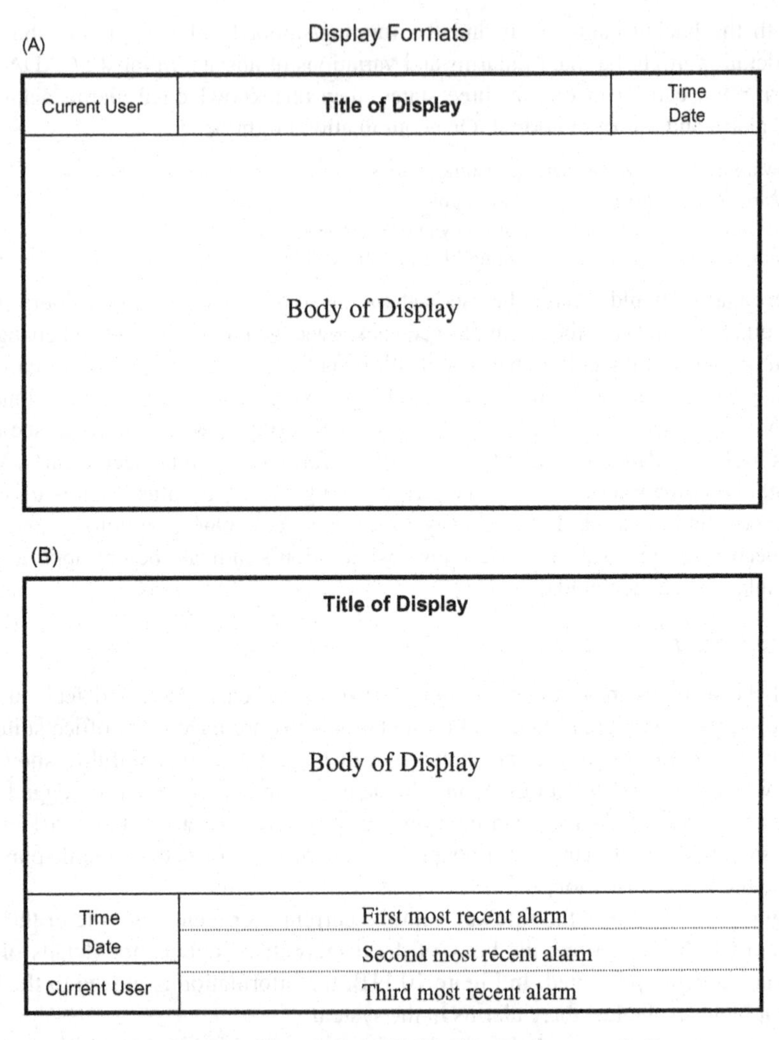

Figure 10.11 (A) Display format example 1. (B) Display format example 2.

A new alarm should make the user aware, such as flashing text or an alarm icon that flashes; sometimes using a standard icon, such as a bell, could be shown in the middle of the display, or perhaps in the title bar at the top of the display. The user would then access the alarm summary display to identify the new alarm condition. An alternative for accessing the alarms would have the user click on the icon to access the current alarm summary. Likewise, if an alarm became 'acknowledged' or 'return to normal', then again there would be the notification. Since alarms and their associated conditions are important, there should be some sort of visual/audible notification when an alarm occurs.

There was one municipality that had designated critical alarms which are of higher priority than all other alarms; if one of these critical alarms occurs, and the operator did not address the condition within a set time period, then the alarm is re-annunciated. Consequently, the high priority alarms cannot be ignored or dismissed.

The topics described in this section highlight some of the overall considerations that one should address before getting into the detailed design of the various displays. Since these decisions, and therefore standards, will be used throughout the system, it is important to set these standards first so that all succeeding work adheres to these standards.

10.4.2 Process Graphic Displays

The process graphic displays provide the user with a 'window' into the processes of the SCADA system; specifically, the user is able to observe everything that is happening in the process areas with respect to the system operations. A SCADA system could have many different displays of various types, each conveying different information to the user, and therefore serving different purposes.

In this subsection, the types of displays and some general guidelines will be covered to help organize and design the displays for a SCADA project. The actual design of each display and the layout is entirely up to the programmer or designer of the system.

The general categories or types of displays to be considered are Current Operations, Equipment Control, Alarm and Event Summaries, Trends and Historical Reports, and Maintenance displays.

Current Process Operations

These displays contain the current operating status of the equipment portrayed in the display. Consider the example Clarifier facility in Figure 10.12 to illustrate what current operating information is included. From this display, the operator can see

- Current flowrate through clarifier
- Water turbidity leaving the clarifier
- Level in the filter inlet chamber
- Running/stopped status of clarifier mixer
- Percent open of the inlet valve
- Control mode of the inlet control valve
- Control mode of the mixer and scraper

In this system, dark blue is used to represent the raw water from the Low Lift Station; the running status uses the traffic light colour convention. The backflush and blowdown valves are shown in blue as these are not controlled by the SCADA system.

Note that at a glance, the operator can tell by the colour if the mixer is running, and what the relative level is in the filter inlet chamber; colour conventions help to make this kind of information clear even from a distance from the workstation.

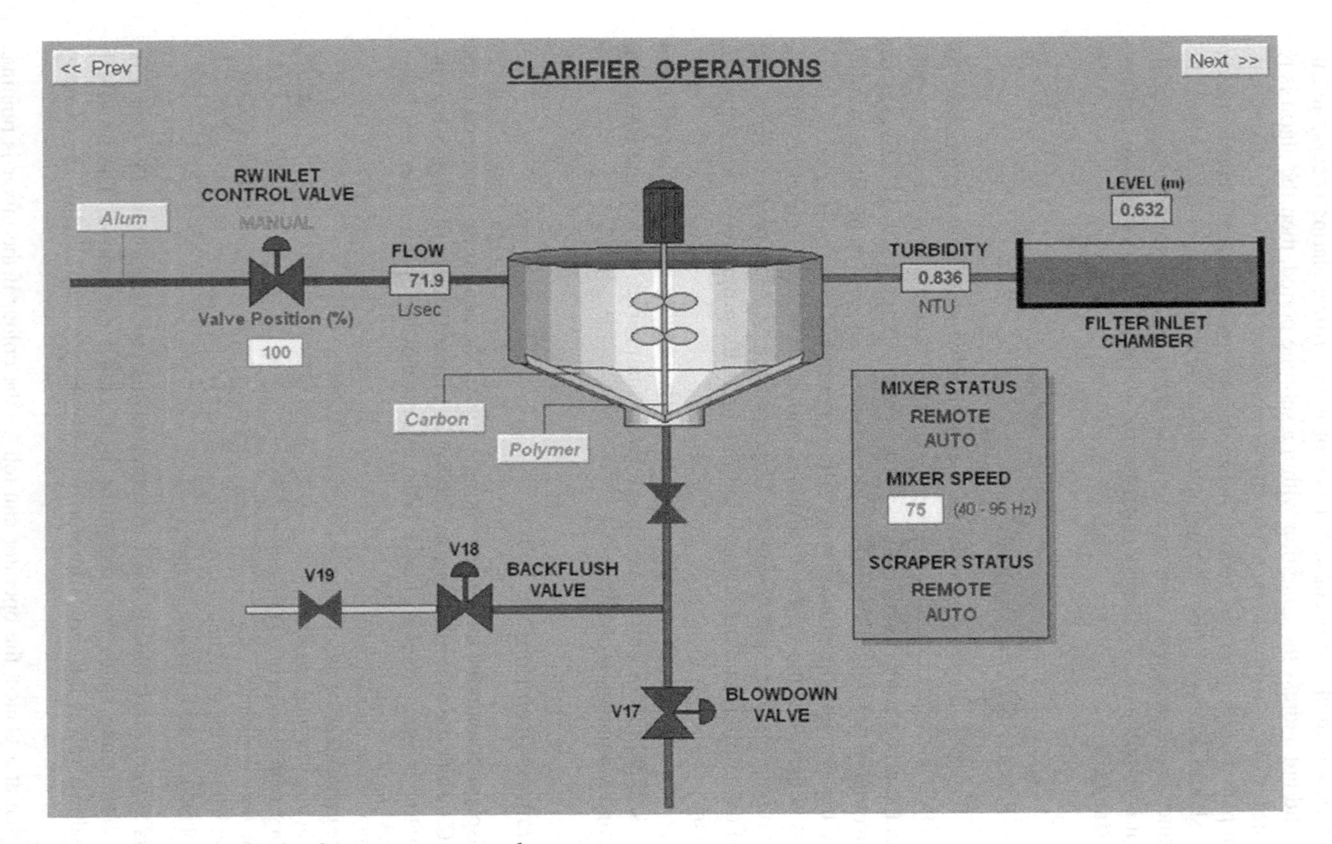

Figure 10.12 Clarifier system display for water treatment plant.

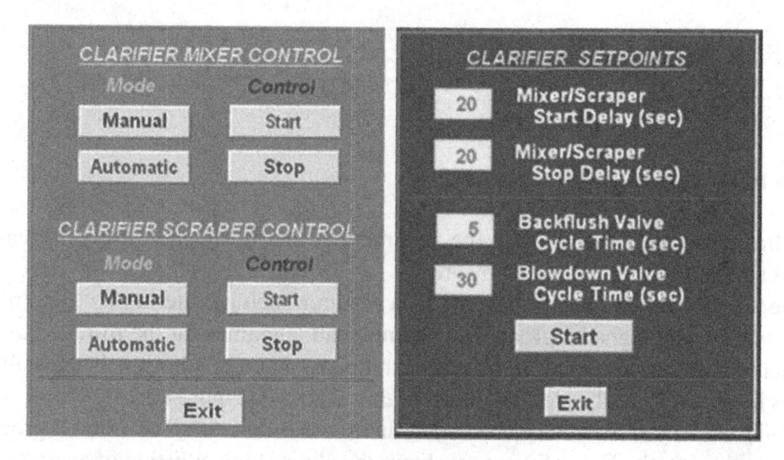

Figure 10.13 Pop-up control panels for Clearwell system.

Equipment Control Displays

The control operations for the equipment has been deliberately left out of this screen to keep the display 'clear' to see. Pop-up displays are used such that clicking on the equipment causes the panels shown in Figure 10.13 to appear. Once the control actions are complete, these displays can be removed using the 'Exit' buttons.

For each controllable device, there are pushbuttons for selecting either 'Manual' or 'Automatic' mode of control; in manual, the operator can use the 'Start' and 'Stop' pushbuttons to control the devices, while in the automatic mode, the control program in the PPC controls the operations. The speed setpoint for the mixer is also shown and can be changed.

Alarm and Event Summaries

The user of the SCADA system needs to know what alarms currently exist, so an Alarm Summary display serves this purpose. All alarm conditions are listed in chronological order, with the date and time of occurrence. SCADA software packages such as InTouch and iFIX include a preformatted alarm summary 'wizard' which can be configured to show a variety of information; typical fields shown include

- Tagname and Description of point
- Date and Time of occurrence
- Current State (discrete) or Value (analog)
- Alarm status – new alarm, return to normal, acknowledged alarm.

Figure 10.14 shows a portion of one such alarm summary for a SCADA system. In this example, both the original time of occurrence, and the most recent time of occurrence are shown; it is possible that an alarm type point could go into alarm, out of alarm and back into alarm state without being acknowledged. The 'Node' field

Ack	Date In	Time In	Time Last	Node	Description	Tagname
✓	5/11/2004	13:42:59.531	13:42:59.531	HCWATER	Low Lift Pump 1 Runtime Exceeded Alarm	LLF_LLP1_RT_ZGA
✓	5/11/2004	13:42:59.531	13:42:59.531	HCWATER	Low Lift Pump 2 Runtime Exceeded Alarm	LLF_LLP2_RT_ZGA
✓	5/11/2004	13:42:59.531	13:42:59.531	HCWATER	Low Lift Pump 3 Runtime Exceeded Alarm	LLF_LLP3_RT_ZGA
✓	5/11/2004	13:42:59.531	13:42:59.531	HCWATER	High Lift Pump 3 Not Auto Alarm	HLF_HLP3_SS_ZGA

Figure 10.14 Sample alarm summary display.

identifies an area of the system; the programmer can designate different areas for alarms so that they can be sorted by area.

The alarm wizards usually have built-in features such as filtering by area (node), date and time and tagname group. For example, all tagnames for the filtration area of the plant could be displayed in the alarm summary. The default view is normally all alarms in chronological order, as shown in this example (Figure 10.14).

While the Alarm Summary shows all alarm conditions or some filtered group of alarm conditions, the Event Summary shows the chronological listing of every change of state of discrete points, and every command change, including setpoint changes. This summary is used to trace a problem in the system; when something fails, it can be helpful to view all events in the system leading up to the failure condition.

For example, a pump might stop unexpectedly, resulting in an 'Uncommanded Stop' alarm and 'Pump Stopped' event. Tracing back through the event summary one might find that the discharge valve on the pump did not open, so the discharge pressure in the pump caused an overload, which in turn caused the motor of the pump to stop. Some failure conditions can be the result of several other events; hence, these event summaries can be useful in troubleshooting a problem.

Figure 10.15 shows a portion of an Event Summary for a SCADA system. For each entry, the date and time of occurrence is shown, along with the tagname, description and event condition. A discrete input such as a valve opening or closing will produce a Change of State, or COS, event; this records the fact that the valve has changed state. Any point in the process database can be designated as Change of State, alarm in the True state, alarm in the False state, or simply an event with no alarm condition associated with it.

In this example, it can be seen that a user has changed a setpoint for a filter; this results in an entry showing the tagname and the new value, together with the date and time of the change.

Trends and Historical Reports

The purpose of trend displays is to view the values of points over a set period of time. The trend display can be configured with several points at once, and the time

```
5/11/2004  13:44:21.4 [HCWATER ] FLT_FIL2_IV_SCD            COS          NotCls    Filter 2 Inlet Valve Close Status
5/11/2004  13:44:26.5 [HCWATER ] FLT_FIL2_IV_SOD            COS            Open    Filter 2 Inlet Valve Open Status
5/11/2004  13:44:27.5 [HCWATER ] FLT_FIL2_EV_SCD            COS          NotCls    Filter 2 Effluent Valve Close Statu
5/11/2004  13:44:29.4 [HCWATER ] FLT_FIL2_EV_SOD            COS            Open    Filter 2 Effluent Valve Open Status
5/11/2004  13:44:30.4 [HCWATER ] FLT_CWL2_IV_SCD            COS          NotCls    Filter 2 Clearwell Inlet Valve Clo:
5/11/2004  13:44:31.4 [HCWATER ] FLT_CWL2_IV_SOD            COS            Open    Filter 2 Clearwell Inlet Valve Oper
5/11/2004  13:44:52.3 [HCWATER ] FLT_FIL2_FC_RSP                             10    Filter 2 Effluent Flowrate Setpoint
5/11/2004  13:44:52.3 [HCWATER ] Fix32.HCWATER.FLT_FIL2_FC_RSP.F_CV set to 10 by HCWATER::KYLE B
5/11/2004  13:45:01.5 [HCWATER ] FLT_FIL2_01_ZPF            COS              On    Filter 2 In Service Mode
5/11/2004  13:45:01.5 [HCWATER ] FLT_FIL2_02_ZPF            COS             Off    Filter 2 Out Of Service Mode
```

Figure 10.15 Excerpt from a sample event summary display.

period can be configured for any time period for which data has been accumulated. The information for the trend displays is retrieved from the historical database and not the process database. This historical database is organized in a compact form to keep the file size as small as possible, while still saving all values saved. As has been stated previously, these historical database files are usually organized as an SQL file. By so doing, other applications can access information from this historical database file.

The historical database can be configured in either of two ways: if points are entered into the process database using the SCADA software package's data entry facility, then the 'archive' or 'historical' check box can be checked to flag the point for historical archiving; if a spreadsheet is used, then the appropriate format for the historical file must be established, and then points can be entered in cells in the CSV file, then imported into the software package. Either way, a point must be declared for archiving by referencing its tagname, and the interval for saving must be specified, such as every 100 ms or every 60 s.

Figure 10.16 illustrates a portion of a trend display in which a number of tank levels are being shown. This trend was created using the trend wizard of the SCADA software package; the points being trended are listed at the bottom, and the value of each point at the vertical cursor is shown in the colour assigned to the point (i.e., the pen colour assigned to the tagname).

Trend displays have a default time period configuration which is set when the trend display is created; however, while viewing, the time period can be changed to zoom in or zoom out as desired. A section of the trend can be selected which is then expanded to fill the display field. The trend shown here is configured for a 12 hour

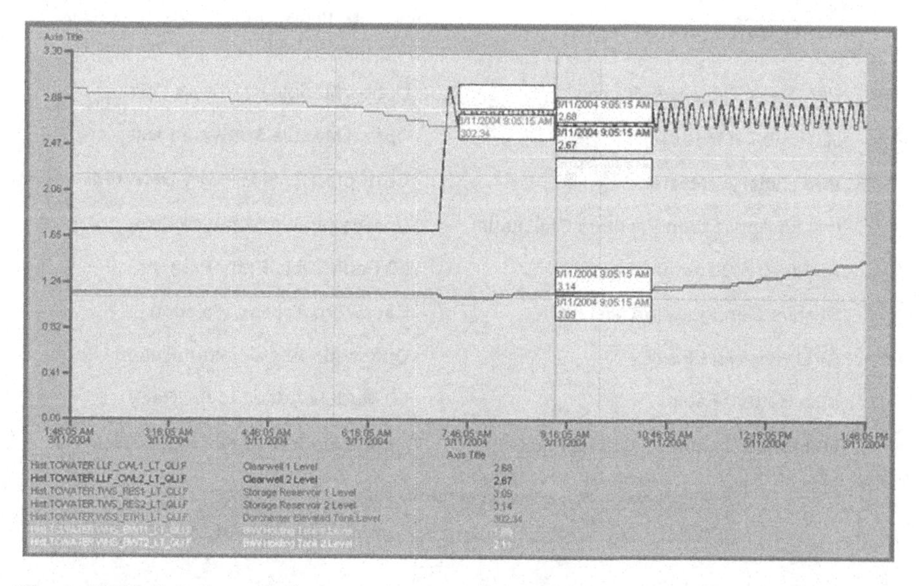

Figure 10.16 Sample trend display.

period, going back from the current time. Note that the points with their tagnames and descriptions are shown with their assigned pen colours.

Maintenance Displays

One final category of displays that are used for troubleshooting and for monitoring the operation of the actual PPCs and communications is the maintenance display: this type of display shows current information for the status of the communication with the PPCs over the plant network, as well as operating information about the PPCs in particular. A maintenance display might show the current state of each PPC (e.g., Running, Faulted, Program), the time-of-day clock values for each PPC and a count of any errors in communications. With today's more advanced PPCs, such as the A-B ControlLogix and the GE Fanuc PACSystem PLCs, there are built-in features which can be used to report problems to the SOWs through displays.

Figure 10.17 shows a template or wizard display from the GE Fanuc Cimplicity HMI software; the information here informs the user of conditions inside the processor of the PLC. This template wizard is for the GE Fanuc 90–30 PLC family. For example, if the hardware configuration is changed, a 'System Configuration Mismatch' alarm will occur; or if the processor detects an error in the application program, then an 'Application Program Fault' alarm will occur. This display illustrates the various alarm and error conditions than can be detected directly by the processor of the PLC.

Other maintenance type displays can be created which retrieve information from the processor of the various PPCs in the system, and then show them on displays like the one illustrated.

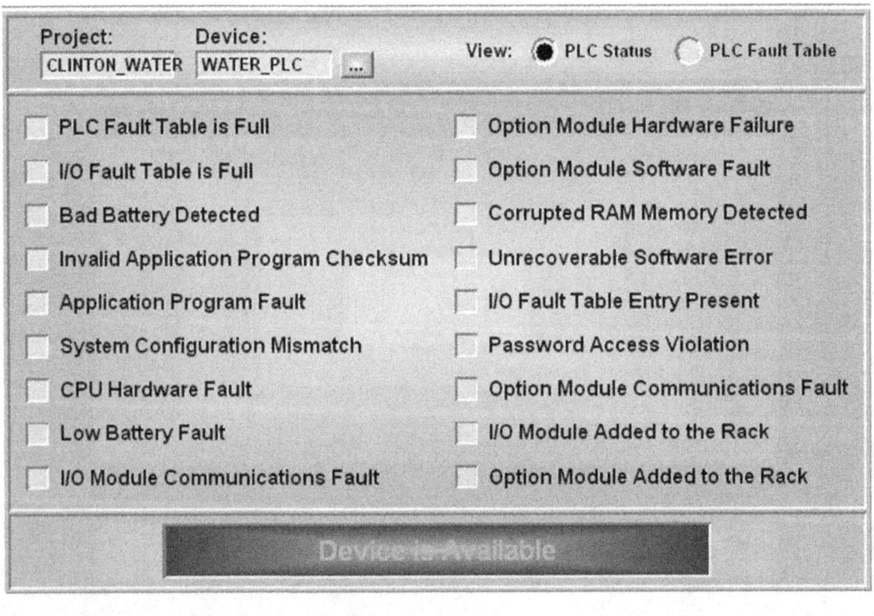

Figure 10.17 PLC diagnostic display.

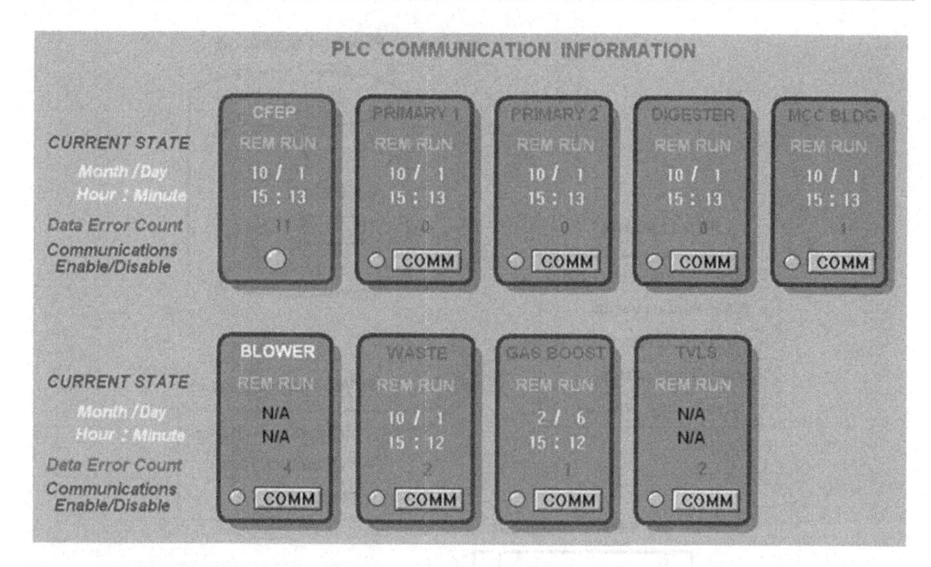

Figure 10.18 PPC monitoring display.

Another type of maintenance display can be configured to show the time-of-day clocks in the PPCs, along with any communication failures that occur. One project that I worked on experienced a number of inter-PPC communication failures; consequently, I created the maintenance display shown in Figure 10.18. In this display, information from each of the PPCs is shown, including the current operating mode (REM RUN for remote run), time-of-day clock, and the number of communication errors. This last data item was needed as the network sometimes became overloaded, and information between PPCs did not get transferred; by monitoring this communication status, a high number of errors would indicate a problem at a particular PPC.

10.5 Configuration of I/O Server

One of the primary functions of the Server workstation in a SCADA system is to communicate with the various PPCs on the network, in order to poll or retrieve current values of points within the PPC application programs. For example, a PPC which monitors and controls the Low Lift Station of a water treatment plant will be polled for the current states of the equipment (i.e., pumps), current values of key analog inputs (i.e., flowrate, reservoir level, water quality parameters) and possibly the current values of some of the internal virtual software points (i.e., total flows, runtime).

10.5.1 Conceptual Operation of the I/O Server

Each point in the process database that references a PPC includes information about the hardware with which to communicate. The point would reference a hardware

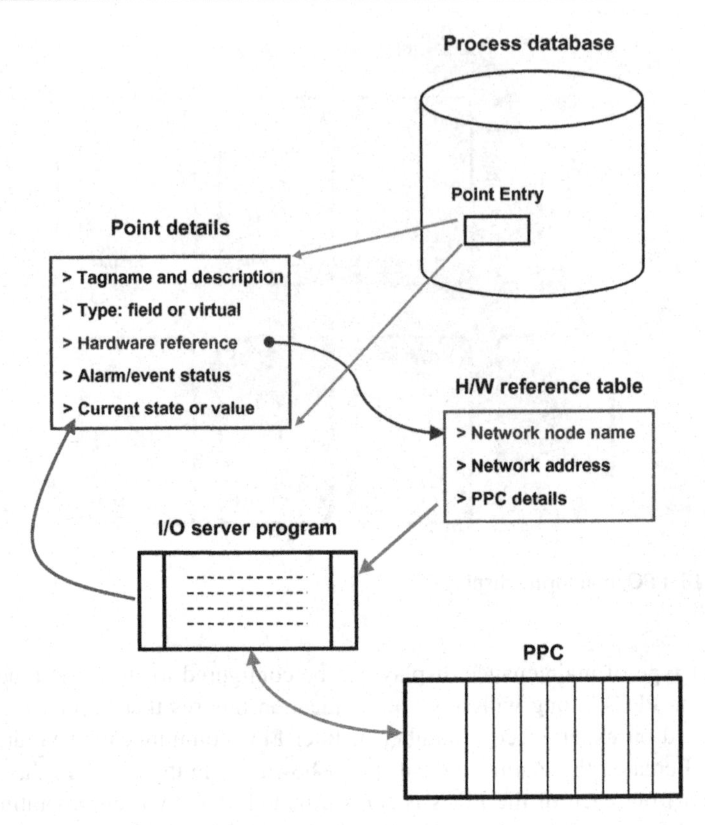

Figure 10.19 I/O server communications illustration.

device, such as the PPC and the address within the PPC. A hardware reference table would contain the connection information between the PPC being referenced and the physical location of the PPC (e.g., IP address or other identification of the PPC on the network).

Figure 10.19 illustrates how the process database entry is linked with the I/O server program to access the data values in the PPCs. An I/O server program would use the information in the Hardware Reference Table and the data to be retrieved to communicate with the PPC to obtain the current value. Once the data value(s) had been received, then the I/O server program would update the current value field of the database point. The I/O server program would be configured with a list of the points to be accessed in the PPCs and the data to be retrieved or polled.

The various SCADA software packages perform this communications differently, but the concept is still the same: a link between each point in the process database identifies where in the network the data resides; specifically, which PPC and what address within the PPC. This allows the I/O server program to communicate with the PPCs to continuously retrieve the current values for the various points in the process database.

10.5.2 Considerations of the Network Topology

In Section 2.2.4, the concept of network topology was introduced; the choice of topology affects the speed with which data points can be polled for values. Here we will consider the Bus and Star configurations, and how they compare in performance.

The *Bus Topology*, as shown in Figure 2.2, uses a single data path for all communications. Today's Ethernet IP communications is very fast, and this arrangement generally works quite well. If a system has a large number of PPC nodes on the network and/or there is a lot of data to be transferred to and from the server, then this configuration could become overloaded. This would result in communication failures and retries on the data transfers. The effect at the SOWs is that the display updates may be slower.

The *Star Topology*, as shown in Figure 2.3, uses multiple paths out from the server, one path for each PPC node. This requires more hardware and is more expensive to implement, but is noticeably faster. Large systems with many PPC nodes would benefit from this arrangement, since each PPC node has its own direct communication path to the server. However, the server could then become overloaded as it tries to communicate with many nodes at the same time.

One solution to the overloaded server is to split the workload into two or more servers; each server workstation would be configured to handle a fixed number of nodes and the process database would actually consist of two or more databases. All of the server workstations would be connected together via a bus network, so the SOWs could access any data point by issuing requests to the appropriate server computer. Having multiple servers does increase the complexity of the system, but for large systems with thousands of data points in the database, this is the better arrangement.

Finally, one must consider the issue of redundancy in the server: since this is the key workstation in terms of maintaining system information, it would be prudent to use dual redundant servers and dual redundant switches. If one server in a pair fails for any reason, then the other server can take over. If one of the switches between the servers and the network failed, then again the alternate switch can perform the work. This redundancy is an important consideration when designing the network topology, since one must ensure that the communication works reliably and there is a backup mechanism in place.

11 System Integration, Commissioning and Checkout

Objectives

- Testing and verifying Software In-House
- Factory Acceptance Testing
- On-Site Field I/O Checkout
- Site Acceptance Testing
- Commissioning Application Software

Today's SCADA systems can vary greatly in size and complexity; a logical and organized approach is needed to test and verify all of the application software developed for the SCADA system. Whether the system consists of one Programmable Process Controller (PPC) and one Operator Terminal (e.g. touch screen interface module) or many PPCs interconnected via a communication network (i.e. Ethernet and TCP/IP) and multiple operator and supervisor workstations, all aspects of the project must be tested, verified and commissioned. Testing and verification should be done both in house (at the factory) and on site; the more testing that can be done in house, the less time will be spent on site. This in-house testing could translate into savings in cost.

11.1 In-House Factory Testing and Checkout

The application software programs for the PPCs should follow the structured approach as described in Chapter 9: using subroutines to organize the program logic will make the testing easier and save time during commissioning. The methods and suggestions in this section are based upon a well-structured project.

11.1.1 Checking Overall Program Flow

As was explained in Chapter 9, today's PPCs which implement the IEC 61131-3 programming standard allow for three types of tasks, each of which can have multiple programs: continuous, synchronous based upon a time period and asynchronous which are triggered by a condition in the PPC. The methodology which follows applies to all three types of tasks.

If the mainline routine of each program consists primarily of subroutine calls to other modules, then software points can be used to enable and disable the calls. The tagnames for these software points could simply be the name of the program

Designing SCADA Application Software. DOI: http://dx.doi.org/10.1016/B978-0-12-417000-1.00011-7

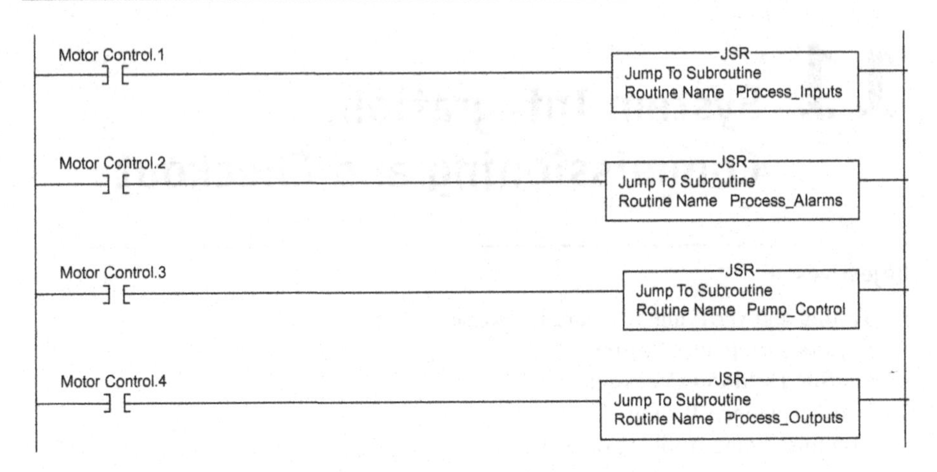

Figure 11.1 Subroutine calls using virtual flags.

followed by a routine number. For example, a program name of Motor_Control could use a Double Integer (i.e. 32 bit field) in which each bit is used as an enable point for a call, Motor_Control.1 would enable the first subroutine call, Motor_Control.2 would enable the second subroutine call and so on. This block of points could be created in each of the program's own tag database, thus keeping each program's flags separate from all others.

Consider some sample logic for the 'Motor_Control' program just mentioned: the logic shown in Figure 11.1 might be used in the mainline routine for this program. The four subroutines created here are being called 'Motor_Control.x', based upon the enabling flag point. With 32 bits available in the double integer data type, a program could have up to 32 subroutines. If more flags are needed, then a second and/or third Double Integer could be created.

To check the basic program flow and to ensure that all of the routines are present, the system programmer could enable the flags one at a time and check the calls to the routines. To verify that the subroutine did actually execute, a simple counter of logic could be temporarily added to each routine, perhaps as the first or second rung. A block of counters could be created similar to the enable flags, in which an array of counters is created for the program; again, these temporary counters could be created within the program's tag database. Consider the following logic for one of the subroutines of a program, as shown in Figure 11.2.

Each time the subroutine is called, its respective counter will increment by one; if the call sequences are working correctly, then one should see that the counters are incrementing steadily, typically at the scan rate of the program. After each routine's call has been verified, the counter rung can be removed. Each subroutine of each program of each task can thus be tested for basic program flow; any problems will become apparent quite quickly, and they can then be rectified.

The unlatch coil in parallel with the up-counter instruction is required to reset the 'count-up' condition that was set True when the subroutine was called. If this flag bit

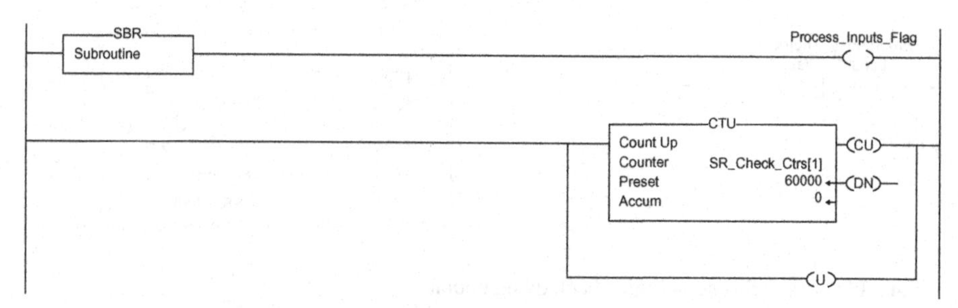

Figure 11.2 Logic to confirm subroutine call.

in the counter is not reset, then the counter will not count further, as the logic state input to the counter will remain True. By resetting the 'count-up' flag, the state is returned to False, ready for the next trigger of the counter in the subroutine.

The preceding description can easily be checked for the main continuous task and its programs; but how can the programs within the synchronous tasks be tested? The same method still applies, except that the rate at which the counters increment will be based upon the interval time set for the parent task.

If a task was configured to execute once every 50 ms, then each routine in each program will see its counter increment 20 times per second (i.e. at 50 ms, the task will execute 20 times per second). So the counter technique still works, only one must now verify that the counters increment the correct number of times in 1 s. If one 'watches' the execution for a period of time, such as 1 min, then the resulting counts should be 120 (60 s of 20 passes).

Finally, the asynchronous task can be checked in a slightly different manner, but still using a counter. Since the event- or interrupt-type task executes once when a programmed condition occurs, then every time that condition is triggered, then the counter within the task's program should increment once. For testing purposes, either a field input signal or a virtual software point could be used as the trigger condition; when the trigger input changes from False to True, then the interrupt task should execute once, and the counter inside the routine of the task's program should increment once. After the interrupt program flow has been verified, then the testing trigger condition can be replaced with the correct trigger condition, knowing that the program flow is correct. The counter test logic can be removed once the invocation has been tested (Figure 11.3).

It is worth noting that the more thoroughly the application software is tested in house, the less time will be spent testing on site. The commissioning of the application software for a SCADA project is always the last part of the project; one cannot test all of the software until all other trades or work have been completed. This testing on site for commissioning purposes uses the live system, so the more off-site testing that can be done, the better the site commissioning will proceed. The unfortunate part of this is that the system programmer will be under great pressure from all other parties to get the project completed, so the pressure on the programmer can be quite intense.

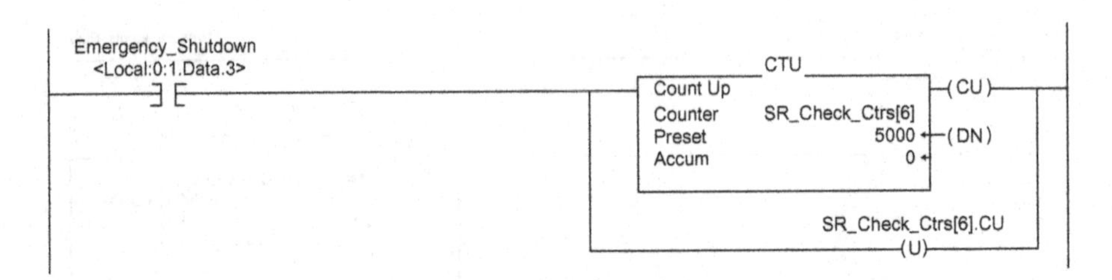

Figure 11.3 Asynchronous logic check using counter.

The other trades involved in the project may be late on the project schedule for any number of reasons, so the overall project timeline slips; in some projects, there are liquidated damages that apply if the project is not 'substantially complete' by a specified date, unless a reasonable explanation can be provided. If there are extras to the contract which require changes and more time, then that is generally agreed to and the schedule is amended. But all too often, one or more parties involved takes longer than expected, and the cumulative effect is that the project is significantly late by the time the programming personnel are able to test their work. So even though the programming work was not late, all the pressure will be on the programmer to complete the project.

Speaking from personal experience, I have been put in this position more than once; by the time I was allowed to test my application software on site, the project was already significantly late, and everyone waited on me. If the liquidated damages clause was applied, then the other parties pointed their collective fingers at me, because I was the last person to commission. Fortunately, I was able to explain and show that I was not to blame for the delays, but it did require that I keep a detailed log of my efforts and especially every time another party caused a delay that would impact my schedule. So as a suggestion from someone who has been there, test as much of the software as possible in house before going to site, as this will save considerable time commissioning, and just might keep you from having damages applied to you, when the delay was not due to you!

11.1.2 Testing Program Logic Using Simulation

Automation systems are often quite large, involving multiple PPCs with complex application programs. One needs a systematic approach to test and verify every part of the project. In this section, a checklist template is suggested as a way of ensuring that every module of every program of every task has been tested.

There are two approaches that can be used: test the logic in each module separately and test subsystems as a whole. Perhaps both methods can be used to perform an incremental test procedure rather than trying to test the entire system all at once. If the latter approach is used, one can get lost in the details and potentially miss or overlook some aspects of the design.

Module Verifications

The module checkout could be done using a checklist template similar to the one shown in Figure 11.4; this checklist is organized to include every software component of the application program for each PPC; note that one or more such checklists would be required for each PPC in the system.

One added benefit of this checklist is that it shows the hierarchy of the programs and tasks, and it ensures that every module of logic is accounted for. Starting from the left column, the names of the tasks, program and included routines are listed in an indented manner. A date field can be used to indicate when the particular module's logic had been verified. The 'Pass' and 'Error' columns can be checked to indicate the result of the module logic testing. In the case of errors, an error log can be maintained to describe the problems encountered and that need to be addressed.

Testing each module requires simulating inputs to the module logic and checking that the results agree with what was expected. For example, if an alarm handling routine includes logic to verify that the motor run status becomes True within an allowed time period, and there is a timer which combines the 'Start' command and the 'Running' status, then two cases can occur: the run status is True within the allowed time, thus disabling the timer and the run status does not become True within the allowed time, thus enabling the timer to complete. In the latter case, an alarm condition should be generated as a result of the timer completing, and setting its 'Done' flag, which in turn, sets a 'Failed To Start' alarm flag or condition.

The logic for this start/stop checking is shown in Figure 11.5.

This logic shows that the Run Command output and the Run Status are combined using timers. If either timer expires before the desired state is reached, then a software alarm flag is set True. The preceding description explains how this logic operates.

Process Verifications

If the individual modules check out, then putting them together into complete subsystems should prove quite a bit simpler since the basic logic has already been proven. The subsystem testing involves identifying each of the processes in the design and then organizing them into another checklist, similar to the one shown in Figure 11.6, in which the various processes are listed.

The task and the program are identified first, followed by each of the processes in the program. The description is for a particular part of the process, and again there are columns for date, pass and errors. The processes listed in this checklist are based upon the design in the Process Control Logic Descriptions (PCLDs); one must verify the operation of the application software against the original design documents. It was noted earlier that complete software documentation is important in these SCADA projects; here is an example of how such documents can be helpful. Since the design has already been reviewed and approved during the design phase, then the program logic must follow that design. Hence, there should be no surprises when the

SOFTWARE MODULE TESTING CHECKLIST

Project: Low Lift Pumping Station

PPC: LLF_PPC
Node ID: 192.168.42.103

Task	Program	Main Routine	Module	Module		Date	Pass	Error
Continuous								
	Primary_Control							
		Control_Main						
			Process_Inputs					
			Process_Alarms					
			Process_Outputs					
				Discrete_Ctrl				
				Analog_Ctrl				
			Motor_Control					
				Control_Modes				
				Motor_Controls				
	Secondary_Control							
		Second_Main						
			Status_Check					
			Controls					
			Interfaces					
Interval_1								
	Data_Collection							
		Data_Main						
			Averages					
			Data_Storage					

Figure 11.4 Sample module testing checklist.

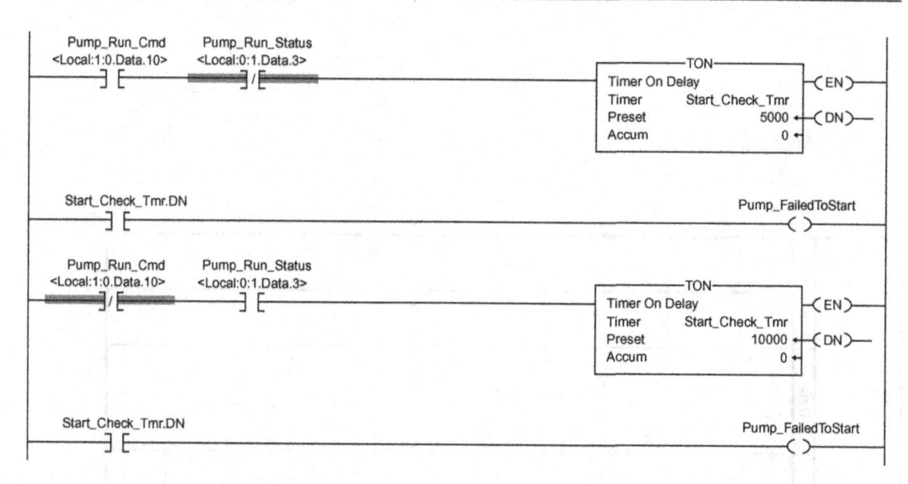

Figure 11.5 Check logic for motor start/stop.

programmer gets to this stage of testing the actual programming logic that has been created.

If there were changes to the design earlier on, then this program testing phase will be based upon the revised design, and the programmer will not have to be concerned about such issues as: is this logic correct and is it based upon the agreed to design? Like all other trades on a project, any change must be submitted for approval before being implemented. In the field of software development, it is tempting to take the attitude 'sure, I will add that in, it is only a small change'. By the end of the project, those 'small changes' can add up to a substantial amount of extra work, none of which is being paid for. So always make sure that any change is documented and approved before proceeding with the detail design and implementation.

11.1.3 Factory Acceptance Testing

Some contracts require a Factory Acceptance Test (FAT) conducted at the contractor's facility; such a test consists of the contractor, or the system programmer in particular, demonstrating that the program logic works correctly. With the PCLDs and PPC points lists in hand, the various processes of the design are simulated and the operations are verified. The test is usually performed with the completed panel, with the PPC and all terminations. Input signals are jumpered to simulate the real world inputs; then the actions of the program logic are verified for correctness.

Typically both the consultant and the customer are present for the testing, so that both parties are satisfied that the system has been implemented correctly. Problems may still show up on site during the commissioning, but at this FAT stage, much of the logic can be checked out and demonstrated.

The FAT is based upon written procedures, usually written by either the consultant or the system programmer; the latter has a more detailed understanding of all the logic involved, so it is not uncommon for him/her to prepare this document.

PROCESS OPERATION TESTING CHECKLIST

Project: Low Lift Pumping Station

PPC: LLF_PPC
Node ID: 192.168.42.103

Task	Program	Process	Component		Date	Pass	Error
Low Lift Station							
	Low_Lift						
		Travelling Screens					
			Manual/Auto control mode selection				
			Start/Stop control of screens				
		Low Lift Pumps					
			Mode and Duty selection				
			Remote Manual control operation				
			Remote Automatic control operation				
	Data_Collection						
		Pump runtime accumulation					
		Flowrate totalizing to flows					

Figure 11.6 Sample process operation testing checklist.

The document itself does not necessarily have to be lengthy or too detailed; usually the key operating characteristics must be listed with some details of their respective operations.

The Process Operation Testing Checklist introduced in the last section could be used as a basis for the FAT. If all of the processes in the design have been included in these checklists, then these lists might serve the FAT very well. The formal FAT would then be essentially a repeat of what was already done, but would be witnessed by and signed off by the customer and the consultant. It is important to get the FAT officially signed off since this represents a major milestone in the overall project. The successful completion of the FAT serves as a benchmark for the system during site commissioning.

11.2 On-Site Field I/O Checkout

The first task in system checkout on site is to verify the wiring of the field I/O signals, from the field device to the terminal strips in the panel to the inputs and outputs of the PPCs. The signals should have been checked between the terminal strips and the PPC I/O modules in the factory, so that the FAT could be completed using simulated inputs to the PPC. The remaining I/O checkout involves verifying that the devices are wired to the correct inputs and outputs, and the signal statuses agree with the original design.

11.2.1 Field I/O Signal Checklists

The field I/O signal points lists that were produced earlier in the project can be used for testing all of the signals in and out of the PPC. Figure 11.7 shows a modified page for discrete inputs in which two more columns have been added: the wired state of the input and a field to enter the date when each signal was verified. By adding the wired state column, the spreadsheet not only indicates the terminal numbers of each input, but also the 'relaxed' state of the input, Normally Open (NO) or Normally Closed (NC). The sample is shown with these new fields completed.

Figure 11.8 shows a similarly modified spreadsheet for analog input signals, but here the new column is used to enter the calibrated range of the signal or instrument. Analog input instruments typically use the 1–5 or the 4–20 mA range; regardless of signal type, the calibration of the signal is important to know, so the extra field here can be used to record this calibration. In this example, engineering range calibrations are shown, along with verification dates.

All of the points lists for all of the PPCs can be modified to add the additional columns; this information can then serve as documentation showing that not only was every input and output verified from the device to and from the PPC module, but that the discrete signal wiring state and the analog signal calibration are recorded.

Water Treatment Plant
Raw Water Pumping Station

Controller Name: RWPSTN
Network IP: 192.168.205.141

Tagname	Point Description	Point Type	Hardware Address	PLC Module Prod. Code	Tag Reference	Terminal Numbers	Wired State	Date Checked
LLF_RWP2-00-SRN	*LL Pump 2 Running Status*	DI	I:7/0	1746-IA8	GB-020	1120-07	N.O.	Apr. 14
LLF_RWP2-00-SCM	*LL Pump 2 Remote Mode*	DI	I:7/1	1746-IA8	HI-020	1120-10	N.O.	Apr. 14
LLF_RWP2-00-AGF	*LL Pump 2 General Fault*	DI	I:7/2	1746-IA8	XS-020	1120-11	N.O.	Apr. 14
LLF_RWP2-PS-APL	*LL Pump 2 Low Pressure Alarm*	DI	I:7/3	1746-IA8	PSL-021	1120-12	N.C.	Apr. 14
LLF_RWP2-PS-APH	*LL Pump 2 High Pressure Alarm*	DI	I:7/4	1746-IA8	PSH-022	1120-13	N.C.	Apr. 14
LLF_RWP2-DV-SCM	*LL Pump 2 DV Remote Mode*	DI	I:7/5	1746-IA8	HI-024	1120-14	N.O.	Apr. 15
LLF_RWP2-DV-SCD	*LL Pump 2 DV Close Status*	DI	I:7/6	1746-IA8	ZSC-024	1120-15	N.O.	Apr. 15
LLF_RWP2-DV-SOD	*LL Pump 2 DV Open Status*	DI	I:7/7	1746-IA8	ZSO-024	1120-16	N.O.	Apr. 15
LLF_RWP3-00-SRN	*LL Pump 3 Running Status*	DI	O:4/0	1746-OA8	GB-060	1130-00	N.O.	Apr. 15
LLF_RWP3-00-SCM	*LL Pump 3 Remote Mode*	DI	O:4/1	1746-OA8	HI-060	1130-01	N.O.	Apr. 15
LLF_RWP3-00-AGF	*LL Pump 3 General Fault*	DI	O:4/2	1746-OA8	XS-060	1130-02	N.O.	Apr. 15
LLF_RWP3-PS-APL	*LL Pump 3 Low Pressure Alarm*	DI	O:4/3	1746-OA8	PSL-061	1130-03	N.C.	Apr. 16
LLF_RWP3-PS-APH	*LL Pump 3 High Pressure Alarm*	DI	O:4/4	1746-OA8	PSH-062	1130-04	N.C.	Apr. 16
LLF_RWP3-DV-SCM	*LL Pump 3 DV Remote Mode*	DI	O:4/5	1746-OA8	HI-064	1130-05	N.O.	Apr. 17
LLF_RWP3-DV-SCD	*LL Pump 3 DV Close Status*	DI	O:4/6	1746-OA8	ZSC-064	1130-06	N.O.	Apr. 17
LLF_RWP3-DV-SOD	*LL Pump 3 DV Open Status*	DI	O:4/7	1746-OA8	ZSO-064	1130-07	N.O.	Apr. 17

Figure 11.7 Discrete input points list checkout sheet.

Water Treatment Plant **Raw Water Pumping Station**		*Controller Name:* RWPSTN *Network IP:* 192.168.205.141						
Tagname	Point Description	Point Type	Hardware Address	PLC Module Prod. Code	Tag Reference	Terminal Numbers	Scaled Range	Date Checked
LLF_RWP1-IE-QII	*LL Pump 1 Current*	AI	I:1.0	1746-NI8	II-010	1160-00	0 - 600 A	Apr. 14
LLF_RWP2-IE-QII	*LL Pump 2 Current*	AI	I:1.1	1746-NI8	II-020	1160-01	0 - 600 A	Apr. 15
LLF_PWL2-LT-QLI	*Low Lift Well 2 Level*	AI	I:1.2	1746-NI8	LIT-050	1160-02	0 - 2.3 m	Apr. 15
LLF_RWP3-IE-QII	*LL Pump 3 Current*	AI	I:1.3	1746-NI8	II-060	1160-03	0 - 600 A	Apr. 15
		AI	I:1.4	1746-NI8		1160-04		
LLF_RWP1-DV-QZI	*LLP1 Discharge Valve Position*	AI	I:1.5	1746-NI8		1160-05	0 - 100 %	Apr. 20
LLF_RWP2-DV-QZI	*LLP2 Discharge Valve Position*	AI	I:1.6	1746-NI8		1160-06	0 - 100 %	Apr. 20
LLF_RWP3-DV-QZI	*LLP3 Discharge Valve Position*	AI	I:1.7	1746-NI8		1160-07	0 - 100 %	Apr. 20
LLF_RWP1-DV-KSP	*LLP1 Discharge Valve Setpoint*	AO	O:2.0	1746-NO4I		1150-00	0 - 100 %	Apr. 21
LLF_RWP2-DV-KSP	*LLP2 Discharge Valve Setpoint*	AO	O:2.1	1746-NO4I		1150-01	0 - 100 %	Apr. 21
LLF_RWP3-DV-KSP	*LLP3 Discharge Valve Setpoint*	AO	O:2.2	1746-NO4I		1150-02	0 - 100 %	Apr. 21
		AO	O:2.3	1746-NO4I		1150-03		

Figure 11.8 Analog input points list checkout sheet.

11.2.2 Verifying Field Signals

The procedure for verifying each of the inputs and outputs for each PPC requires the electrician to energize each input signal, and then the programmer verifies that the correct state is observed in the program logic. For example, if a motor run contact is wired NO, then when the electrician closes the contact, the programmer should see an energized contact in the program code.

For analog input signals, a calibrated signal generator is used by the electrician to inject a known signal into the field side; the programmer should then see the correct engineered scaled value in the program logic. For a 4–20 mA input, a 12 mA signal injected for a point which is calibrated as 0–5000 rpm, the programmer should observe a value of 2500 rpm in the program logic.

On the output side, the programmer would force each discrete output to the True state, and the electrician would verify that the field state is True; perhaps a discrete output drives a field relay, so the energized state in the logic would result in the relay being energized. For analog output signals, the programmer would again force an engineered scaled value in the logic, and the electrician would measure the output signal; for a VFD scaled in per cent, for example, setting the speed output in the code to 50% should result in the field signal indicating 12 mA (half of 4–20 mA).

This process is repeated for every input signal to the PPC. As each signal is tested and verified, the additional information is added to the two columns shown in the example in Figures 11.7 and 11.8.

11.2.3 Correlating Field Wiring and Logic States

For discrete input points, the most common error or problem is to have the wired state and the logic state reversed. An alarm type input is often wired as a NC contact, so that the failure of the signal will be to 'fail open', and thus create an alarm. There may be situations in which these alarm signals were wired as NO contacts, which in turn may require inverting the type of instruction in the program (check for closed state vs check for open state). In the A-B Programmable Logic Controllers, an Examine If Closed instruction may have to be replaced with an Examine If Open instruction.

Figure 11.9 shows a comparison of the wired state of an input switch, the NO or NC instruction in the logic, and the resulting logic state that will result. For example, for an NC connection on a switch, such as the Emergency Stop switch, an NO instruction will result in a True logic state when the Emergency Stop pushbutton is at rest; that is the switch has not been pressed. If this switch is pushed in, indicating a stop condition, then the resulting logic state will become False.

Consider the second group shown for the NO connection: for a motor run contact, which is normally wired as an NO contact, then the NO instruction will result in a True logic state only when the motor is running. Likewise, a Local/Remote selector

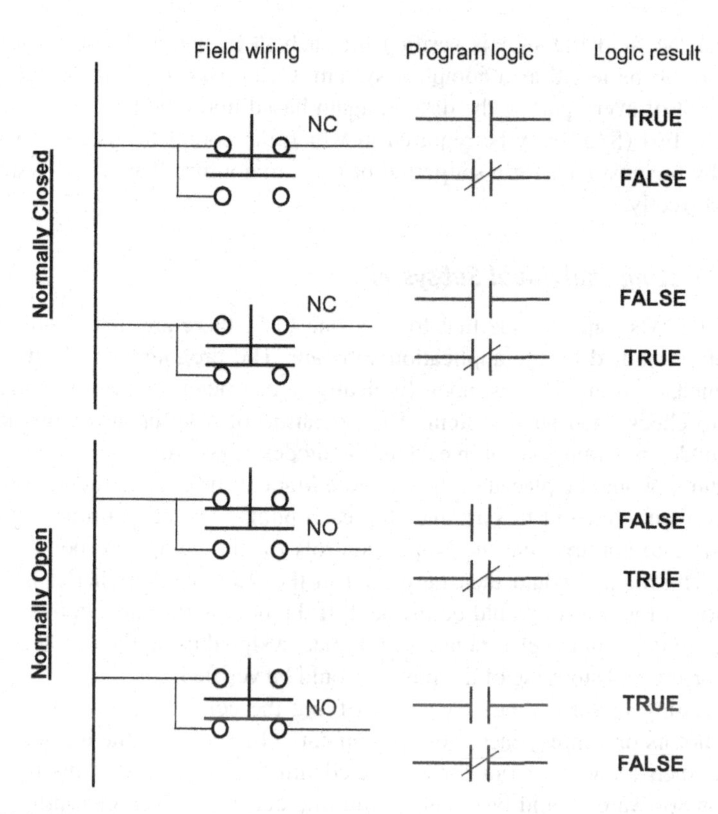

Figure 11.9 NC and NO wiring examples.

switch in the Remote position will energize the electrical state to True, so an NO instruction will result in a True logic state.

It is important that the field wiring agrees with the contract drawings, but at the same time, the program logic must work correctly. I have had some situations on projects in which the wiring was correct, but the state as seen in the software was inverted; so either the wiring had to be switched or the software had to be switched. One of the benefits of programming is that changing from an NO instruction to an NC instruction takes a few seconds, whereas changing the wiring could take 3–5 min. One such situation involved about five or six Local/Remote selector switches for some chemical systems, in which the Local state on the switch enabled my logic. To save time, I reversed the instructions rather than having the electrician rewire the switches; the result was that it took 30 s to change the software instead of 20 min to change the wiring.

11.3 Commissioning and Site Acceptance Testing

With all of the field I/O signals verified for each PPC, the application software is finally ready to be tested as a complete system. Commissioning involves a systematic checkout of every part of the design, again based upon the PCLDs. A final Site Acceptance Test (SAT) may be required in which the complete system is used and operated by the customer for a set period of time, to confirm that all subsystems are working correctly.

11.3.1 Testing Individual Subsystems

From the PCLDs, one can see that for any one PPC, there are many separate processes being handled by the application software. The programmer, electrician and a representative from the customer (typically a maintenance person) must work together to check each such system. The operation of a group of pumps must be operated under program control in each of the modes of control.

Each pump would be placed in the 'Remote Manual' mode, and the customer representative would attempt to start and stop each pump. The programmer would follow the logic to confirm that the output controls for the pumps are being activated correctly. The pumps would then be placed in the 'Remote Automatic' mode, and the conditions for control would be invoked. If the pumps start and stop based upon an analog signal, then the electrician would inject a signal using the signal generator, and the starting and stopping of the pumps would be verified.

Some testing requires time, such as verifying the correct accumulation of runtimes for motors or pumps. Each motor/pump could be started in the Remote Manual mode and then allowed to run for a selected time period. During this period, the application software should be accumulating the correct amount of runtime. At the end of the set time, such as 30 min or 2 hours, the accumulated time in the software would be checked against the time period for which the equipment was left to operate. Assuming that the times match, then the accumulation of runtimes has been verified for the software.

This type of step-by-step testing would proceed through each of the processes defined in the design of the software, until all processes have been verified. It might prove useful to have a copy of the PCLDs on which each part is 'checked off' as that subsystem or process has been checked. A date could be entered next to the sections to record when each portion had been verified.

11.3.2 Site Acceptance Testing

Some projects may require a formal testing procedure on site to verify that not only do the individual processes work correctly, but that the entire system operates without any major problems or failures for some predefined period of time. In the days of minicomputer systems, crashes and other problems might occur even after all of the operations had been tested separately. Consequently, there were often formal SATs required which identified what had to happen over what period of time, before the works were accepted by the end user and the consultant. With today's software systems, in which the hardware and software being used have a proven record, this type of test may not be required.

11.3.3 Final Project Sign Off

Once all of the processes have been tested for all of the PPCs, and the functionality of the SCADA user workstations has been proven, there is often a testing period in which the end customer uses the system for a set period exercising the system as much as possible. The user or customer will use the system, testing all the possible subsystems to ensure that everything functions as it was intended. Sometimes, an issue will arise in which the customer believes some part is not operating as it should. One must return to the detailed design documentation to determine if the customer's expected results match the design documents. Once again, having detailed design documents from the beginning becomes very important, as no one wants to go back and redesign and reprogram any part of the system.

Once the system passes this last testing phase, all documentation is updated to include any final changes and adjustments that were implemented. During the final commissioning phase, it is not unusual to make some changes and leave the documentation updates until later; at this stage of the project, the important thing is to get the project completed successfully. However, one should certainly make detailed notes on any changes so that the final documentation edits can be done quickly and efficiently.

The final documentation, as described throughout this book, would then be submitted and approved; at this point, the application programming is done!

12 Sample Project – Applying the Principles

Objectives

- Apply principles of book to programming project
- Illustrate procedures as applied to a real-world project

Having presented all of the principles that I believe should be included in the design and development of a SCADA system application, this chapter presents an example programming project for a water pumping station. This application covers the design and programming of the Programmable Process Controller (PPC) application program and the design of the SCADA Operations Workstation (SOW) graphic displays and database.

The information and illustrations included in this chapter will be limited, as to include the entire project would be too long for this book. Since the PPC is at the heart of any SCADA system, the focus will be on the controller program. Specific detailed examples from each step of the process will be included so as to illustrate the application of the principles and ideas presented in this book.

12.1 Overview of Project

The water pumping station application program involves the operation of and the control for two pumps, which involves both Manual and Automatic modes of control, and the collection of data for use at the operator workstation. The programming is based upon the Allen-Bradley ControlLogix PLC (Programmable Logic Controller), as this PLC is a well-established controller in the field of automation systems. The A-B ControlLogix PLC serves as a good example for demonstrating structured programming techniques.

This chapter illustrates the application of all of the principles described in the book, guidelines and methods described in this book, to create a real-world application program for a known PLC. As will be seen in this chapter, each of the steps outlined in Chapter 3 will be shown as applied to this project.

Designing SCADA Application Software. DOI: http://dx.doi.org/10.1016/B978-0-12-417000-1.00012-9

12.2 Identify Process Area Field Signals

The first step in defining the field signals is to identify the equipment involved, and from that, define the inputs and outputs required for the application. This application includes two pumps, an intake well from which water is drawn, an effluent control valve, and a reservoir level signal which is used for the control of the pumps in the automatic mode of control.

The field signals for this application are listed below, grouped according to the type of signal: discrete input, discrete output, analog input and analog output. As with any application program, there are virtual or internal software points required for the application program in the Programmable Process Controller. These will be shown later in this section.

Two Pumps:	DI	running status
	DI	control mode (local/remote)
	DI	general fault
	DI	DV open status
	DI	DV closed status
	DO	pump run control
	DO	DV open control
	DO	DV close control
Effluent Valve:	DI	discharge valve remote select
	AI	discharge valve position
	AO	discharge valve control
Intake Well:	DI	high level alarm
	DI	low level alarm
Analog Signals:	AI	reservoir level
	AI	station discharge flowrate
	AI	chlorine residual
	AI	water turbidity
	AI	water temperature
	AI	water pH.

12.3 Create and Document Application Databases

The starting point for creating the database is to use the points list templates described previously in Chapter 6, and then populate these points lists with the various field I/O points and internal software points identified for the project. Both the controller application program and the workstation application require a database which can be created by saving spreadsheets in CSV format, and then importing them into the application software system.

12.3.1 Developing the Controller Database

The field I/O points list serves as the basis for the controller database; these points represent the physical inputs and outputs. Organizing these signals into points lists and assigning module types, the spreadsheets shown in Figure 12.1 represent the

(A)

Remote Puming Station **Treated Water Pumping Station**		RemotePumpStation *Sample Pumping Station Points List* Page: 1			
Tagname	Point Description	Point Type	Hardware Address	PLC Module Prod. Code	Terminal Numbers
RPS_BSP1_PB_SSP	*PS Pump 1 Stop Pushbutton*	DI	Local:1:I.Data.0	1756-IA16	
RPS_BSP1_PB_SST	*PS Pump 1 Start Pushbutton*	DI	Local:1:I.Data.1	1756-IA16	
RPS_BPS1_00_AGF	*PS Pump 1 General Fault*	DI	Local:1:I.Data.2	1756-IA16	
		DI	Local:1:I.Data.3	1756-IA16	
RPS_BSP1_DV_SCD	*PS Pump 1 DV Closed Status*	DI	Local:1:I.Data.4	1756-IA16	
RPS_BSP1_DV_SOD	*PS Pump 1 DV Open Status*	DI	Local:1:I.Data.5	1756-IA16	
RPS_BSP1_00_SRN	*PS Pump 1 Run Status*	DI	Local:1:I.Data.6	1756-IA16	
RPS_BPS1_SS_SCM	*PS Pump 1 Control Mode*	DI	Local:1:I.Data.7	1756-IA16	
		DI	Local:1:I.Data.8	1756-IA16	
		DI	Local:1:I.Data.9	1756-IA16	
		DI	Local:1:I.Data.10	1756-IA16	
		DI	Local:1:I.Data.11	1756-IA16	
		DI	Local:1:I.Data.12	1756-IA16	
		DI	Local:1:I.Data.13	1756-IA16	
		DI	Local:1:I.Data.14	1756-IA16	
		DI	Local:1:I.Data.15	1756-IA16	

Figure 12.1 (A–E) Pumping Station points lists.

(B)

Remote Puming Station
Treated Water Pumping Station

RemotePumpStation
Sample Pumping Station Points List

Page: 2

Tagname	Point Description	Point Type	Hardware Address	PLC Module Prod. Code	Terminal Numbers
RPS_BSP2_PB_SSP	*PS Pump 2 Stop Pushbutton*	DI	Local:2:I.Data.0	1756-IA16	
RPS_BSP2_PB_SST	*PS Pump 2 Start Pushbutton*	DI	Local:2:I.Data.1	1756-IA16	
RPS_BPS2_00_AGF	*PS Pump 2 General Fault*	DI	Local:2:I.Data.2	1756-IA16	
		DI	Local:2:I.Data.3	1756-IA16	
RPS_BSP2_DV_SCD	*PS Pump 2 DV Closed Status*	DI	Local:2:I.Data.4	1756-IA16	
RPS_BSP2_DV_SOD	*PS Pump 2 DV Open Status*	DI	Local:2:I.Data.5	1756-IA16	
RPS_BSP2_00_SRN	*PS Pump 2 Run Status*	DI	Local:2:I.Data.6	1756-IA16	
RPS_BPS2_SS_SCM	*PS Pump 2 Control Mode*	DI	Local:2:I.Data.7	1756-IA16	
		DI	Local:2:I.Data.8	1756-IA16	
		DI	Local:2:I.Data.9	1756-IA16	
		DI	Local:2:I.Data.10	1756-IA16	
		DI	Local:2:I.Data.11	1756-IA16	
		DI	Local:2:I.Data.12	1756-IA16	
RPS_WWL1_LS_ALH	*PS Intake Well High Level Alarm*	DI	Local:2:I.Data.13	1756-IA16	
RPS_WWL1_LS_ALL	*PS Intake Well Low Level Alarm*	DI	Local:2:I.Data.14	1756-IA16	
RPS_PSD1_EV_SCM	*PS Discharge Valve Remote*	DI	Local:2:I.Data.15	1756-IA16	

Figure 12.1 (Continued)

(C)

Remote Puming Station
Treated Water Pumping Station

RemotePumpStation
Sample Pumping Station Points List

Page: 3

Tagname	Point Description	Point Type	Hardware Address	PLC Module Prod. Code	Terminal Numbers
RPS_BSP1_00_DRN	*PS Pump 1 Run Control*	DO	Local:3:O.Data.0	1756-OA16	
RPS_BSP1_DV_DON	*PS Pump 1 DV Open Control*	DO	Local:3:O.Data.1	1756-OA16	
RPS_BSP1_DV_DCE	*PS Pump 1 DV Close Control*	DO	Local:3:O.Data.2	1756-OA16	
		DO	Local:3:O.Data.3	1756-OA16	
		DO	Local:3:O.Data.4	1756-OA16	
		DO	Local:3:O.Data.5	1756-OA16	
		DO	Local:3:O.Data.6	1756-OA16	
		DO	Local:3:O.Data.7	1756-OA16	
RPS_BSP2_00_DRN	*PS Pump 2 Run Control*	DO	Local:3:O.Data.8	1756-OA16	
RPS_BSP2_DV_DON	*PS Pump 2 DV Open Control*	DO	Local:3:O.Data.9	1756-OA16	
RPS_BSP2_DV_DCE	*PS Pump 2 DV Close Control*	DO	Local:3:O.Data.10	1756-OA16	
		DO	Local:3:O.Data.11	1756-OA16	
RPS_RES1_01_DAI	*Reservoir High High Level Alarm*	DO	Local:3:O.Data.12	1756-OA16	
RPS_RES1_02_DAI	*Reservoir High Level Alarm*	DO	Local:3:O.Data.13	1756-OA16	
RPS_RES1_03_DAI	*Reservoir Low Level Alarm*	DO	Local:3:O.Data.14	1756-OA16	
RPS_RES1_04_DAI	*Reservoir Low Low Level Alarm*	DO	Local:3:O.Data.15	1756-OA16	

Figure 12.1 (Continued)

(D)

Remote Puming Station Treated Water Pumping Station		RemotePumpStation *Sample Pumping Station Points List*			Page: 4
Tagname	Point Description	Point Type	Hardware Address	PLC Module Prod. Code	Terminal Numbers
RPS_RES1_LT_QLI	*PS Reservoir Level*	AI	Local:5:I.Ch0Data	1756-IF6I	
RPS_BSP0_FT_QFI	*PS Station Discharge Flowrate*	AI	Local:5:I.Ch1Data	1756-IF6I	
RPS_PSD1_EV_QZI	*PS Discharge Valve Position*	AI	Local:5:I.Ch2Data	1756-IF6I	
		AI	Local:5:I.Ch3Data	1756-IF6I	
		AI	Local:5:I.Ch4Data	1756-IF6I	
		AI	Local:5:I.Ch5Data	1756-IF6I	
RPS_INL1_AN_QRC	*PS Water Chlorine Residual*	AI	Local:6:I.Ch0Data	1756-IF6I	
RPS_INL1_AN_QNI	*PS Water Turbidity*	AI	Local:6:I.Ch1Data	1756-IF6I	
RPS_INL1_AN_QTI	*PS Water Water Temperature*	AI	Local:6:I.Ch2Data	1756-IF6I	
RPS_INL1_AN_QHI	*PS Water pH*	AI	Local:6:I.Ch3Data	1756-IF6I	
		AI	Local:6:I.Ch4Data	1756-IF6I	
		AI	Local:6:I.Ch5Data	1756-IF6I	

Figure 12.1 (Continued)

(E)

Remote Puming Station
Treated Water Pumping Station

RemotePumpStation
Sample Pumping Station Points List

Page: 5

Tagname	Point Description	Point Type	Hardware Address	PLC Module Prod. Code	Terminal Numbers
RPS_PSD1_EV_KSP	*PS Discharge Valve Position Control*	AO	Local:7:O.Ch0Data	1756-OF6VI	
		AO	Local:7:O.Ch1Data	1756-OF6VI	
		AO	Local:7:O.Ch2Data	1756-OF6VI	
		AO	Local:7:O.Ch3Data	1756-OF6VI	
		AO	Local:7:O.Ch4Data	1756-OF6VI	
		AO	Local:7:O.Ch5Data	1756-OF6VI	
			Ethernet Interface	1756-ENBT	

Figure 12.1 (Continued)

(A)

Water Treatment Plant Raw Water Pumping Station		Page: 11	
Tagname	Description	Base / Alias	Point Type
RPS_BSP1_01_YRN	*Pump 1 Manual Run Control*	Base	Boolean
RPS_BSP1_02_YRN	*Pump 1 Automatic Run Control*	Base	Boolean
RPS_BSP1_DV_YCE	*Pump 1 DV Close Control*	Base	Boolean
RPS_BSP1_DV_YON	*Pump 1 DV Open Control*	Base	Boolean
RPS_BSP1_PB_XST	*Pump 1 HMI Start*	Base	Boolean
RPS_BSP1_PB_XSP	*Pump 1 HMI Stop*	Base	Boolean
RPS_BSP1_PB_XRA	*Pump 1 Remote Automatic Select*	Base	Boolean
RPS_BSP1_PB_XRM	*Pump 1 Remote Manual Select*	Base	Boolean
RPS_BSP1_PF_YAU	*Pump 1 Lead Duty Automatic Select*	Base	Boolean
RPS_BSP1_PF_YDS	*Pump 1 Lead Duty*	Base	Boolean
RPS_BSP1_PF_ZGA	*Pump 1 Software General Alarm*	Base	Boolean
RPS_BSP1_PF_ZRA	*Pump 1 Automatic Mode Selected*	Base	Boolean
RPS_BSP1_PF_ZRM	*Pump 1 Manual Mode Selected*	Base	Boolean
RPS_BSP2_01_YRN	*Pump 2 Manual Run Control*	Base	Boolean
RPS_BSP2_02_YRN	*Pump 2 Automatic Run Control*	Base	Boolean
RPS_BSP2_DV_YCE	*Pump 2 DV Close Control*	Base	Boolean
RPS_BSP2_DV_YON	*Pump 2 DV Open Control*	Base	Boolean
RPS_BSP2_PB_XST	*Pump 2 HMI Start*	Base	Boolean
RPS_BSP2_PB_XSP	*Pump 2 HMI Stop*	Base	Boolean
RPS_BSP2_PB_XRA	*Pump 2 Remote Automatic Select*	Base	Boolean
RPS_BSP2_PB_XRM	*Pump 2 Remote Manual Select*	Base	Boolean
RPS_BSP2_PF_YAU	*Pump 2 Lead Duty Automatic Select*	Base	Boolean
RPS_BSP2_PF_YDS	*Pump 2 Lead Duty*	Base	Boolean
RPS_BSP2_PF_ZGA	*Pump 2 Software General Alarm*	Base	Boolean
RPS_BSP2_PF_ZRA	*Pump 2 Automatic Mode Selected*	Base	Boolean
RPS_BSP2_PF_ZRM	*Pump 2 Manual Mode Selected*	Base	Boolean

Figure 12.2 (A and B) Software points lists.

field I/O points list for the pumping station; the module assignments are based upon the ControlLogix controller.

Software points are required for interfacing with the user via the SOW. The software or virtual points interact with the program logic in the PPC program. Software points for the Human–Machine Interface (HMI) are used to select the mode of control as well as start and stop the pumps in manual mode. Other points are used for setpoints used in the program.

Figure 12.2 illustrates the software points used in this application program. In these pages, the type of point is shown, such as Boolean, Real, Timer and Counter. Tagnames with square brackets represent arrays of points, such as three counters for the runtime accumulation, and six timers for the verification of the start/stop and open/close operations.

(B)

Water Treatment Plant Raw Water Pumping Station			Page: 12
Tagname	Description	Base / Alias	Point Type
RPS_PSD1_EV_KSP	*Station Effluent Valve Control*	Base	Real
RPS_RES1_LS_ZLI	*Reservoir Low Low Level*	Base	Boolean
RPS_RES1_LS_ZLL	*Reservoir Low Level*	Base	Boolean
RPS_BSP1_Runtime[3]	*Pump 1 Runtime*	Base	Counter
RPS_BSP1_Timers[6]	*Pump 1 Check Timers*	Base	Timer
RPS_BSP2_Runtime[3]	*Pump 2 Runtime*	Base	Counter
RPS_BSP2_Timers[6]	*Pump 2 Check Timers*	Base	Timer
HMI_BSP0_CF_VDS	*Current Lead Duty Pump*	Base	Dint
HMI_BSP1_01_NPV	*Pump Duty 1 Start Level Setpoint*	Base	Real
HMI_BSP1_02_NPV	*Pump Duty 1 Stop Level Setpoint*	Base	Real
HMI_BSP1_03_NPV	*Pump Duty 2 Start Level Setpoint*	Base	Real
HMI_BSP1_04_NPV	*Pump Duty 2 Stop Level Setpoint*	Base	Real
HMI_PSD1_EV_VSP	*HMI Effluent Valve Control Setpoint*	Base	Real
HMI_RES1_01_NPV	*Reservoir High High Level Setpoint*	Base	Real
HMI_RES1_02_NPV	*Reservoir High Level Setpoint*	Base	Real
HMI_RES1_03_NPV	*Reservoir Low Level Setpoint*	Base	Real
HMI_RES1_04_NPV	*Reservoir Low Low Level Setpoint*	Base	Real
RPS_RES0_LS_YLH	*Virtual Reservoir Level High*	Base	Boolean
RPS_RES1_LS_ALH	*RS Reservoir High Level Alarm*	Base	Boolean
RPS_RES1_LS_ALL	*RS Reservoir Low Level Alarm*	Base	Boolean
RPS_RES1_LS_ZLH	*Reservoir High Level*	Base	Boolean
RPS_RES1_LS_ZLM	*Reservoir High High Level*	Base	Boolean

Figure 12.2 (Continued)

From the process graphic displays, for example, the user would click on the Remote Manual pushbutton which would activate or set True the 'RPS_BSP1_PB_XRM' tag; in the logic, this would cause the Pump 1 Remote Manual mode selected point to become True: 'RPS_BSP1_PF_ZRM'. The other software points operate in a similar manner, setting the modes of control and invoking operations for the pumps and effluent valve.

12.3.2 Developing the SCADA Workstation Database

Since the operations workstation is intended to reflect the controller in the SCADA system, the database points from the PPC can be used to create the workstation database. Using the points list from the controller, one can import these points into HMI software application. Section 6.4 described the procedure for using the controller points list database to create the workstation database. For the sake of brevity, a simplified explanation will be offered here, with the details being offered in Chapter 6.

First, a few sample database points would be created within the HMI software application, typically one of each point type. For example, a discrete input, discrete output, analog input and analog output points would be created. This collection of points would be exported to a Comma Separated Value (CSV) file and saved. Refer to Figures 6.23 and 6.24 for examples of these initial database entries.

Second, the controller points list for the sample pumping station would be placed into a CSV file, for importing into the workstation software application. The controller points list spreadsheets would be placed into a CSV file; the entries in the controller CSV file would then be copied and pasted into the workstation CSV file, using the copy and paste feature to fill in the other columns of data. The populated CSV file would then be imported back into the workstation application, and thus create the completed database. Figure 6.25 illustrates part of the CSV after populating the CSV file with the discrete points from the controller.

The software points lists shown in Figure 12.2 illustrates the software points that would be included in the SOW database. These points would also be added to the CSV file for the workstation database. The completed CSV file would then be imported back into the HMI application to create the completed process database.

12.4 Defining and Documenting Controller Operations

The Process Control Logic Descriptions (PCLDs) must be created for this PPC application program before the detailed logic code can be developed. It is the PCLD which provides the details of the application design, which in turn, is used to generate the program logic for the PPC. The application must be divided into the different process operations, and the details of each operation must be defined. As described in Chapter 7, the application processes should be documented into four major sections:

- System Control Strategy Overview
- Facilities and Parameters
- Control Logic Descriptions
- Special Considerations

12.4.1 Process Summary and Overview

The starting point is to identify the processes being performed by this application, and the equipment involved. Following is a summary of the equipment and the functionality for this application:

- Two pumps operating in Remote Manual and Remote Automatic modes
- Effluent valve operating in Remote Manual mode
- Reservoir level signal used to start and stop the pumps in automatic mode
- Level in the reservoir is to be maintained between a low and high limit
- Accumulation of runtime for each pump.

For this application example, the following major process operations have been defined:

- Lead Duty Pump Selection
- Pump Control Mode Selection
- Pump Control Operations
- Effluent Valve Operation
- Pump Runtime Accumulation

Some details associated with these processes, such as verifying that pumps start within the allowed times, will be addressed in the major sections. Since Appendix C contains a complete PCLD for a process area, this chapter will offer an abbreviated version, highlighting the important aspects.

12.4.2 Detailed Process Descriptions

Having identified and described the information for the PCLD we can now organize this material and expand it into a logical description, following the format described previously. Rather than include an entire PCLD here, the description which follows highlights the important points about the content of the descriptions; all of the important details about the PCLD will be shown herein.

System Control Strategy Overview

The sample pumping station program includes logic to monitor and control two pumps and one effluent modulating valve, as well as to monitor process signals such as flowrate and reservoir level. The purpose is to draw water from an intake well and pump it out into the distribution system, maintaining the level of water in a reservoir located downstream of this pumping station. The downstream reservoir maintains system pressure as well as provides a source of water in that area of the system. The level signal is telemetered back to the controller, and is connected to one of the analog inputs of the PPC.

Two pumps operate in either Automatic mode, in which one pump designated the lead duty is used to discharge water to the reservoir, or Manual mode, in which the operator can start and stop each pump from the SOW. The control modes can be selected from the SOW, along with the pump to be designated as the lead duty for automatic control mode.

The pump starts when the level drops below the low level setpoint, and stops when the level rises above the high level setpoint. Additional low low and high high level setpoints are defined for alarming purposes. These setpoints are entered by the operator through the SOW via the process graphic displays.

The effluent discharge valve can be modulated by the operator through the SOW, including setting the valve to 100% open, which would be the normal mode of operation. In emergency situations, it may be desirable to close this valve completely.

The process signals for flowrate, reservoir level, water quality and so on are continuously being monitored. These signals can be saved in the historical database at the SOW for creating trend type displays. With the flowrate signal, the operator can adjust the effluent flow control valve to maintain the discharge flowrate within set limits. One option in logic would be to control the effluent valve with a PID control loop based upon a flowrate setpoint entered by the operator.

Facilities and Parameters

The pumping station includes an intake well into which water from the treatment plant enters. Two pumps are used to draw water from this well and discharge the water out into the distribution system, and into a downstream reservoir. The level in the reservoir is used to start and stop pumps in the automatic mode of control.

The reservoir level signal is compared with high and low level setpoints, which indicate when the pump is to be started and stopped. Operations such as starting and stopping pumps, and opening and closing discharge valves, are timed; setpoints from the SOW are used to verify that these operations are completed within an allowed time.

The key signals used in this application consist of the following:

- Reservoir level
- Operator level setpoints: High High, High, Low, Low Low
- Operator time setpoints for verifying start/stop and open/close operations
- Effluent flowrate for adjusting the effluent control valve.

Control Logic Descriptions

In Section 12.4.1, the main process operations were identified; each of these is described in more detail below.

Lead Duty Pump Selection

One of the two pumps would be selected as the lead duty at any one time. The user at the SOW would enter either a '1' or a '2' into the Lead Duty Pump Number through the workstation, and this value would be used in the application program. The value being either '1' or '2' would set one of the 'Lead Duty Selected' flags in the logic, thus enabling one of the pumps for automatic control.

Pump Control Mode Selection

Both pumps can be operated in Remote Manual and Remote Automatic modes of control. In the former mode, the operator can start and stop the pumps through pushbuttons on the process graphic displays. The operator has complete control over the pumps in this mode.

In the latter mode, Remote Automatic, whichever pump has been designated the lead duty pump will be started and stopped based upon the level in the reservoir, which is being filled by the operation of the pumps in this sample pumping station.

These mode selections cause software points to be set and cleared to indicate the mode of control for each pump. These conditions are then used in the control logic for the pumps.

Pump Control Operations

For the Remote Manual mode, the pumps can be controlled by the operator via the process graphic displays. The start and stop pushbuttons on the graphic displays serve as the controls for the pumps in this mode.

For the Remote Automatic mode, the operator enters level setpoints for the High High, High, Low, and Low Low level conditions for the reservoir. Comparisons are made between the current reservoir level and each of the four level setpoints. When the reservoir level reaches the Low Level, then the duty pump is started; and when the reservoir level reaches the High Level, then the duty pump is stopped.

When each pump is started, its respective discharge valve is opened; and when each pump is stopped, the discharge valve is closed. Timers are used to verify that the pumps start and stop within the allowed time, and timers are also used to verify that the discharge valves open and close within the allowed time. These allowed times are setpoints entered by the operator via the SOW graphic displays.

Effluent Valve Operation

The pumping station includes an effluent valve which can be modulated by the operator through the SOW. There is no automatic mode of control, only manual mode. If the effluent valve is in the Remote mode, then the position setpoint entered by the operator is passed on to the analog output which controls the valve's position.

While this valve can be adjusted by the operator, it is more likely that the valve would be left in the 100% open state. One option here is to control the valve automatically using a flowrate setpoint entered by the operator; a closed-loop PID would be used to control the position of the valve so as to maintain a constant effluent flowrate from the pumping station.

Pump Runtime Accumulation

One of the most commonly collected information in systems is the accumulated runtimes for equipment, in this case, the runtimes for the two pumps. The logic accumulates runtime in hours and tenths of hours.

A periodic or interval task and program are used for this operation. This task executes every 100 ms, or 10 times per second. Each time the program executes, a counter is used to count ten tenths of a second; at this point, a runtime tenths of hours counter is incremented until it accumulates a tenth of an hour, or 6 min. After ten tenths of an hour, the runtime hours is incremented.

A preset for the hours counter can be used to generate an alarm to indicate that the pump requires servicing. For example, after 4000 hours and the counter expires, an alarm can be generated.

Special Considerations

One interlock condition is the Low level condition of the pumping station Inlet Well Level; there are float signals for both high level and low level in this well. If the low level condition exists, then whichever pump is operating in the automatic mode will be stopped, as this condition is used as an interlock in the program logic. An alarm would be generated to the operator to indicate this situation.

For every operation in a SCADA system, there should be some form of check to confirm that the operation completed correctly. In the case of starting and stopping the pumps, check timers are used to confirm that the run status matches the run control condition within an allowed time period; failure here would generate an alarm to the operator via the SOW. Likewise for the operation of the pump discharge valves, in that timers would be used to confirm the opening and closing occurs within allowable time periods.

12.5 Develop Process Controller Application Software

Section 9.3 described the approach to organizing a PPC application program into a well-defined structured application. Today's controllers allow for continuous, interval and interrupt type tasks, and the tasks may have multiple programs and routines. As the complexity of the application increases, the number of programs and tasks would also increase. It is important to remember to use the appropriate type of task and program structure for a given application.

12.5.1 Developing the Overall Project Structure

This pumping station application is relatively small and simple, so a basic continuous task with one program should be sufficient. The one task would have one program, and this program would have a mainline routine which then calls subroutines to handle the various operations. Following is a list of the suggested routines to be included, and the purposes are as follows:

Main Routine	mainline routine which calls other routines
General Functions	general operations associated with entire application
Process Inputs	handle discrete and analog input signals
Alarms	generate alarms for failures and special conditions
Automatic Control	logic for operating the pumps based upon reservoir level
Pump Control	logic for controlling the pumps in each mode
Process Outputs	handle discrete and analog output signals
Effluent Valve	logic to operate the effluent valve in manual mode.

The pump runtimes require maintaining counters of hours and tenths of hours of operation. Additional subroutines could be added to the main program, but another method could be used to ensure accurate update of runtimes. An interval or periodic type task could be configured with one program, whose routine would update the runtimes at consistent intervals of time. Every 100ms, or ten times per second, this program would execute to update the runtime counters based upon the run statuses of the pumps.

Data Collect	accumulate runtimes in hours and tenths of hours.

12.5.2 Developing the Application Program Logic

With the PCLD and the field I/O points lists, the detailed programming of this application can be started. A new project would be created with one continuous task and one interval or periodic task. The continuous task would have one program with multiple routines as described previously; the interval task would have one program with one routine as desribed previously. The detailed program logic could then be created routine by routine. Chapter 9 offers suggestions on ways to create and perform incremental testing of the logic.

As part of the project development, the I/O configuration of modules would be defined; the points lists were shown previously, indicating the choice of I/O modules and the location of each of the field signals for those modules. The A-B ControlLogix PPC, for example, would require this configuration to be completed within the Controller Organizer part of the project. The analog input and output modules would have to be configured for the different signals, such as the reservoir level signal in percent, the effluent flowrate in litres per second and so on.

The final PPC program for this application is shown in its entirety in Appendix D. The listings show all of the routines used in the final program; the various process operations are distributed through those routines as described above.

12.6 Develop SCADA Workstation Application Software

The SCADA application software, often referred to as the HMI software, provides animated graphic displays which inform the user/operator of the current status of equipment and allow for the user/operator to invoke commands through the PPC. In Section 12.3.2, the development of the process database for the workstation was described: import the PPC database points into the application as the basis for the data used in the graphic displays. Once this is done, and the database has been created, one can proceed to design and develop the process graphic displays for this application.

12.6.1 Developing the Process Graphic Displays

For this application involving only one PPC, a single detailed graphic display will suffice; however, an alarm summary type of display would prove useful for viewing any signals in alarm state. Before defining the layout and content of the process graphic display, some conventions must be established as described in Section 10.4 on the development of the workstation application. The following topics are addressed:

- Layout and Standards
- Proposed Graphic Display
- Graphic Animation and Data
- Control Actions

Layout and Standards

First the general layout of the display should be defined, such as the one shown in Figure 10.9; using the version in Figure 10.9B, both the process graphic display

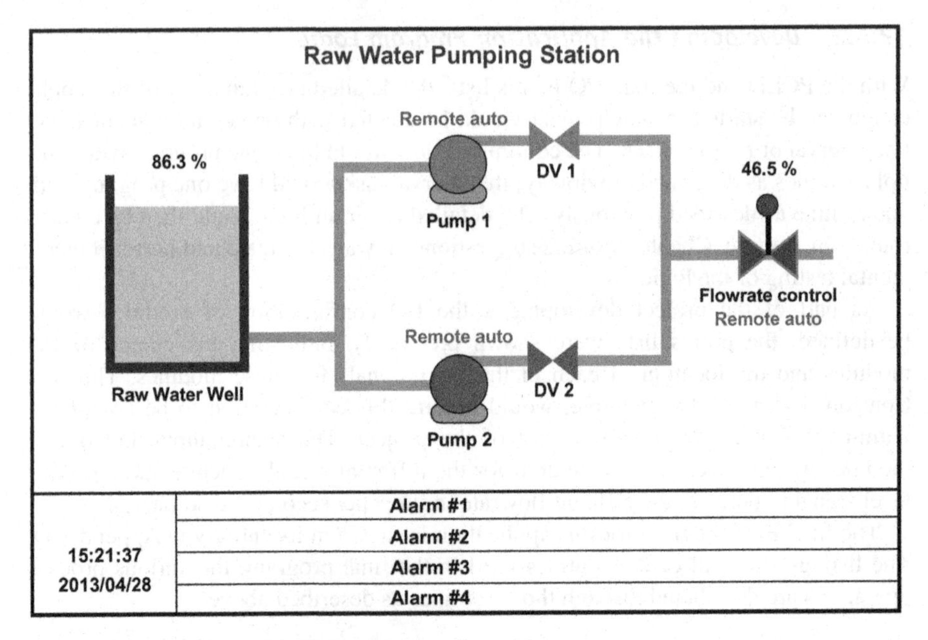

Figure 12.3 Process graphic display for pumping station.

and a three-line alarm can be accommodated. By allocating some real-estate for the alarm window, both requirements can be met.

Next the colour convention should be established: traffic light convention or electrical convention? The former uses green for running (go) and red for stopped (stop), while the latter uses red for running (hot) and green for stopped (ready). I have noticed a trend away from the electrical convention and toward the traffic light convention, since this colour model is better known. With green and red for running and stopped, respectively, a colour for the alarms should be selected. Yellow would be logical, since this fits in with the traffic light concept: yellow means caution. Perhaps flashing yellow could represent a new unacknowledged alarm, and solid yellow could represent an acknowledged alarm. A Return To Normal (RTN) signal could be cyan or blue, which indicates a signal which has returned to a stable or normal state.

Proposed Graphic Display

With these standards established, the detailed process graphic display can be defined. Using the format shown in Figure 10.9B, the process graphic display might appear as shown in Figure 12.3 with the alarms at the bottom.

Graphic Animation and Data

The equipment in the display must then be animated, or linked to the process database points. This display would have the following links or animation:

Raw Water Well:	Numeric value of level
	Animated graphic fill of well

Pumps:	Colour denoting running/stopped status Alarm indicator such as yellow circle inside pump Control mode text using visibility for Manual and Automatic Lead duty text using visibility for lead or blank
Discharge Valves:	Colour denoting open/closed status
Effluent Valve:	Numeric value of percentage open (0–100) Control mode text using visibility for Rem or Loc.

Control Actions

The operator must be able to select the Manual and Automatic modes of control, as well as start and stop the pumps in manual mode; and the effluent valve must be adjustable when in Remote mode. The most common method of implementing controls, particularly on a small application as in this chapter, is to use 'pop-up' windows with the controls inside.

For example, clicking on either pump would cause a control panel to appear overlaid on the process graphic display like that of Figure 12.4; there would be two such panels, one per pump.

This panel contains pushbuttons for selecting Manual and Automatic modes of control, as well as pushbuttons for starting and stopping the pump. The animation links within the pushbuttons would depend upon the SCADA software package; typically, there would be a script which would execute, to issue a command to the PPC. These scripts would be configured to connect with the HMI virtual points in the PPC application software, such as the following:

Pushbutton	PPC Tag
Manual	RPS_BSP1_PB_XRM is toggled True then False
Auto	RPS_BSP1_PB_XRA is toggled True then False
Start	RPS_BSP1_PB_XST is toggled True then False
Stop	RPS_BSP1_PB_XSP is toggled True then False
Exit	Close pop-up control panel.

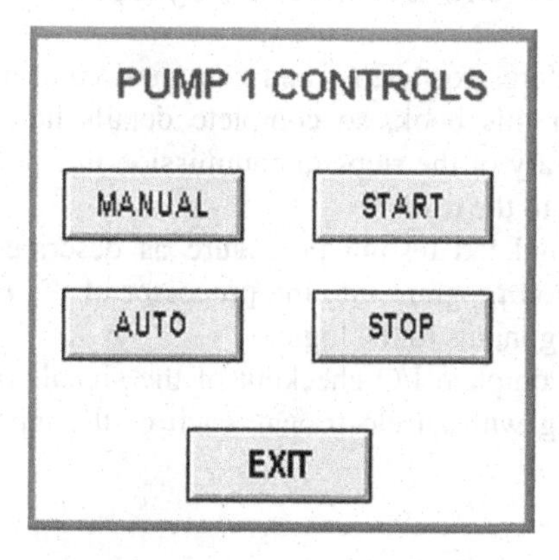

Figure 12.4 Proposed pop-up control panel.

When these virtual points are set True in the PPC control program, the logic would cause the pump to change modes or change state. The toggling of these virtual points would invoke the respective actions in the program.

The effluent control valve would have a similar pop-up control panel to adjust the position of the valve when the valve is in the Remote mode of control.

12.6.2 Configuring the I/O Server Communications

As has been explained previously, the I/O server program or software within the SCADA application software is responsible for communicating with the PPCs of a system. In this application, there is only one such PPC, so setting up the communications is relatively simple.

The points created in the process database must be polled or otherwise retrieved from the PPC on a regular basis; as an example, the data might be retrieved 2–5 times per second. Commands from the workstation to the PPC are configured with some form of scripting feature; for example, when the pushbutton for 'Start' is pressed on the display, a script is executed which toggles on and off the point associated with the virtual start contact in the PPC program. A data change is issued to the PPC to the appropriate point in the database, so that commands to the PPC are executed immediately, while data updates are completed at regular intervals.

Depending upon the SCADA software design, the communications may use an interface such as DDE, NetDDE or an OPC server. I have developed applications in which different groups of points could be polled at different rates, so that key data would be updated quickly, while slower data such as flowrates would be updated less frequently. With today's software systems, the exact configuration of the server depends entirely on how it is designed; hence, no more detailed information can be provided on this topic in this chapter.

Suffice it to say, the I/O server would be configured in some way to retrieve data from the PPC at regular intervals, and commands issued would be sent immediately.

12.7 Commission the Completed System

The methods or procedures for testing and commissioning of a system have been described elsewhere in this book, so complete details here would be repetitive. However, a brief summary of the steps to commission this pumping station application might prove useful to the reader.

First, perform a simulated testing procedure as described in Section 11.1.2 . Download the application program into the processor of the PPC and test all of the operations by simulating inputs to the logic.

Second, perform a complete I/O checkout of the signals on site as described in Section 11.2.1; working with an electrician, each of the inputs and outputs of the PPC can be verified.

Third, proceed with the stepwise testing of each subsystem as described in Section 11.3, to verify the correct operation of all logic.

Fourth, during the testing of the PPC control program, the operation of the process graphic display can be verified by checking that the correct colours and visible text appear to match the intended design. For example, when a pump is in the running state, the colour of the pump in the display should be that of the chosen colour for running. When the mode of control is switched between Manual and Automatic, the corresponding text should appear next to the pump.

These procedures as describe herein and explained in more detail in Chapter 11 have been done by me many times for many projects; I know this approach works well. There may well be other methods of testing and commissioning a system, but the important point is that everything is checked out in a systematic way; once the commissioning is complete, one should be confident that all aspects of the system have been verified to work correctly as designed.

12.8 Finalize Software Documentation

This book has emphasized the importance of good software documentation; if the approach and methods described have been followed, then the finalization of the documentation should prove to be a simple documentation update. It is hoped that the documentation has been kept up to date as the project proceeded, but there are often little details that do not get included in the documents. For this sample project, these details could be noted and at this time, be added to the appropriate document.

In summary, this application for the pumping station should include the following documents:

- Database points lists for both the PPC and the SOW
- Process Control Logic Descriptions for the controller logic
- SCADA User Reference Manual, describing the workstation functions
- Commented program listings for the PPC program
- Design documentation for the SCADA application software.

Appendix A: Glossary of Technical Terms

Appendix A offers a summary of acronyms and terms used throughout this book. Most of these terms are common in the SCADA systems industry.

Alarm Condition in the field in which a fault or an undesirable situation has occurred; requires operator intervention.

Alias Tag Typically the hardware reference for a program point in a controller program; can also refer to another point in the controller program, serving as an indirect reference.

Analog Data Information that can have any number of values between two limits; the limits define the range of the analog value.

ADC Analog to Digital Converter; Electronic interface which monitors a current or voltage and produces a binary number which represents the analog data value; result is stored in binary format in PLC memory.

Asynchronous Task Software task within controller capable of executing multiple programs upon a software Change Of State, such as an input turning on or off; also referred to as an event task.

Binary Data Information stored as '0' and '1'; groups of binary digits (bits) represent data.

Bit Single binary data item, representing one piece of information; BInary digiT.

Byte Group of eight (8) bits, handled as a unit; ASCII characters are represented as one byte per character.

Communication Interface Hardware and software interface modules in PLC which allow data transfer between the PLC and another intelligent device (e.g. PLC, computer and modem).

Continuous Task Software task within controller capable of executing multiple programs in a time-slice manner; programs execute sequentially, start to end, then repeats.

CSV Format Comma Separated Value data, in which each piece of information is separated from the next by a comma; this format allows data to be imported into spreadsheets and other programs for analysis.

Cyclic Task Software task within controller capable of executing multiple programs in a time-slice manner; task and its programs execute one time at regular intervals; used for executing logic at set time intervals; same as the synchronous or event task.

Data Acquisition Process of collecting information from intelligent devices, such as PLCs; typically blocks of data are collected at one time.

Data Archiving Storing of information into historical files for later retrieval into reports; data is normally compacted to save space in files.

Data Highway Common communication path for the transfer of information among all devices connected to the highway, e.g. DH+, Genius Bus and RS-485.

Discrete Data Another term for binary data; discrete data represented by '0' or '1', which corresponds to the two states of the data.

Double Integer Binary representation of an analog value, typically in 32 bits; this represents a number in the range of −2,147,483,648 to +2,147,483,647; allows for larger whole numbers than single Integer.

DPU Distributed Processing Unit; another name for Programmable Logic Controller and Remote Terminal Unit.

Ethernet Collection of networking technologies for interconnecting devices for the transfer of information among the devices on a Local Area Network; high-speed transfers between controllers and workstations.

Event Change in field conditions which represents change of state, such as pump changes from stopped to running status or high-level float switch is tripped.

Event Task Software task within controller capable of executing multiple programs upon a software Change Of State, such as an input turning on or off; same as asynchronous task.

FAT Factory Acceptance Test; series of tests intended to verify the correct operation of a system while in house, prior to shipping to customer site; inputs are simulated to test program logic.

Field Signal Analog or discrete/digital data originating from a field device and is conveyed back to an input module in a PLC; data thus received is then processed through the SCADA system.

Function Block Programming Programming language using graphic blocks with inputs and outputs defined; blocks can be interconnected to create a complete program; high-level graphical language.

Floating Point Representation in binary data form for very large and very small numbers, with great accuracy; follows IEEE floating point format using 32 bits.

Flow Control Algorithm of control in a process in which the rate of flow is controlled using a PLC control program.

GUI Graphical User Interface; interface for humans to interact with software systems using icons, images and graphical images to effect desired actions; start a program within Windows, start a device from an icon.

Graphic Animation Display of active information in the SCADA system; information is changing and updating in real time.
Colour animation – change in colour to denote various states
Graphic animation – change in height or width of object
Numeric animation – change in numeric value of field signal
Visible animation – appearance and disappearance of object.

Historical Data Accumulated information stored with date and time stamping; saved for later retrieval into reports and trends.

HMI Human Machine Interface – previously known as Man–Machine Interface (MMI); process graphic displays and interface used to view and access the SCADA system.

Hyperlink An element or 'hot spot' in a document or display which, when clicked, causes that display to appear in place of the current display; menu displays use hyperlinks to 'jump' to other displays in the system.

Human Machine Interface Represents all of the process graphic and other display information presented to the user via a display monitor.

IEC International Electrotechnical Commission; organization which develops and establishes standards in the technical community, e.g. 61131-3 for programming standards.

IEC 61131-3 Programming standard established by the IEC to define five programming languages and the data types and syntax for programming environment.

ISA Instrument Society of Automation; organization which establishes standards in instrumentation and control systems.

Integer Binary representation of an analog value, typically in 16 bits; this represents a whole number in the range of 0 to 65,535 unsigned or −32,768 to +32,767 signed. Most systems handle long or double integers which use 32 bits instead of 16 bits for more accuracy in arithmetic operations; refer to Double Integer.

Ladder Logic Programming Programming language using graphical rungs resembling electrical wiring diagrams; symbols and instructions execute left-to-right, with input-type operations from the left and output-type operations on the right.

LAN Local Area Network – interconnection of devices including workstations, PLCs, RTUs, printers and telecommunication equipment. Allows for the transfer of information among devices connected to the LAN.

Level Control Control algorithm, based upon maintaining a set level in a storage vessel. Most commonly, level control used to maintain the level of water in an elevated tank, reservoir or standpipe. The PLC program performs all operations to start and stop pumps so as to keep the level within set boundary limits.

Modem Modulator–Demodulator – telecommunication device for converting binary information from a workstation, PLC or other device into a series of tones which are then sent over a communication medium. Typically, two tones are used, one for a binary ‘0’ and one for a binary ‘1’. The medium can be telephone lines, wireless connections, satellite, etc.

Network Bridge Device employing hardware and software which allows transfer of binary information between any two LANs. For example, a bridge may be used to interconnect a SCADA LAN with an Administrative LAN.

Node A device (e.g. computer and printer) that is attached to a computer network such as Ethernet or other telecommunications systems.

OIT Operator Interface Terminal; small panel for users to view and enter information; limited functionality but useful for viewing status of specific equipment and entering/changing setpoints for operations.

Periodic Task Software task within controller capable of executing multiple programs in a time-slice manner; task and its programs execute one time at regular intervals; used for executing logic at set time intervals; same as the synchronous or cyclic task.

PLC Programmable Logic Controller – performs both monitor and control operations, based upon pre-programmed instructions. Provides field control operations and interfaces with workstations executing HMI software.

PLC Interface Special purpose interface module for interconnecting the PLC with other equipment, e.g. other devices using different protocols, workstations and telecommunication systems.

Pop-Up Display Small display window which shows control pushbuttons and/or analog setpoint fields; appears when triggered by a hyperlink in a display and includes an ‘exit’ button to close.

PPC Programmable Process Controller; generic term for field controllers used in automation systems; general term for PLC, RTU, DPU and RPU.

Pressure Control Control algorithm that is based upon maintaining a set pressure in a system. Most commonly, pressure control used to maintain the discharge pressure from a High Lift station or a pumping station. The PLC program performs all operations to start and stop pumps, including varying the pump speeds, so as to keep the pressure within set boundary limits.

Protocol The rules, data formats and data organization used for a specific device. For example, A-B PLCs use the ‘DF1’ protocol, while the Modicon and other PLCs use the ‘Modbus’ protocol. Typically, interface modules are required when interconnecting one device to another when their respective protocols are different.

Quick Link Panel Grouping of pushbuttons on a display for accessing commonly used displays; buttons for alarms, overview, trends, security, etc.; this panel would appear on every display in the workstation screen.

Report-by-Exception Information is transferred from the PLC to the host workstation whenever there is a significant event as defined by the PLC control program. Such events include Change of State of a field signal, an analog value exceeding an alarm setpoint, etc. This method reduces traffic between PLC and workstation by only sending information when there is a change, rather than on a continuous basis as in polling of the PLCs in the system.

Router A device which forwards data packets along networks between any two LANs.

RPU Remote Processing Unit; another name for Programmable Logic Controller or Remote Terminal Unit.

RTU Remote Terminal Unit – originally designed as a PLC for remote locations and communications via telephone lines. Today, the RTU and PLC are essentially synonymous.

SAT Site Acceptance Test; series of tests intended to verify the correct operation of a system as part of the final commissioning of a project; real world inputs and outputs are used to test program logic.

SCADA Supervisory Control and Data Acquisition – Automation system for collecting and analyzing the data from multiple remote devices (i.e. PLCs and RTUs) while providing remote control of field processes through workstations.

Setpoint/Limit Analog value which identifies a boundary of limit for a process value. Setpoints may be used to determine when pumps are to be started and stopped, as well as determine when a field signal is in or out of alarm condition.

SOW SCADA Operations Workstation; generic term for computer workstation executing SCADA HMI software; provides user interface to automation system to view and monitor operations and to effect control over field equipment.

SQL Structured Query Language; special language for entering and accessing the data in a Relational Database Management System (RDBMS).

Structured Text Programming language using high-level programming constructs such as If-Then-Else, Do-While and Case; similar to languages such as 'C' and 'Java'.

Sequential Function Chart Programming language using blocks (steps) and diamonds (decisions) in a graphical form to illustrate program logic as a series of steps, each step being one operation; resembles conventional flowcharts, for simplifying logic flow.

Synchronous Task Software task within controller capable of executing multiple programs in a time-slice manner; task and its programs execute one time at regular intervals; used for executing logic at set time intervals; same as the periodic or event task.

Tagname Alphanumeric text string to assign name to point in application software; applies to field controller and HMI programming; details of point associated with tagname, such as hardware address, description and data type.

TCP/IP Transmission Control Protocol/Internet Protocol; collection of protocols for transmitting data over a network, such as Ethernet.

TSNC Tagname Signal Naming Convention; structured alphanumeric tagname system for naming points in controller and HMI workstation software; uses 16-character tagnames which are structured into fragments or parts of a tagname, with each fragment providing further detail into the nature and purpose of the point.

Appendix B: TSNC Dictionaries

Appendix B presents the details of the tagname system presented in this book, with complete 'dictionaries' of the fragments used in each part of the tagnames. The tagname fragments are based upon the many water and wastewater system projects that I have developed.

B.1 Tagname Signal Naming Convention

The TSNC, or Tagname Signal Naming Convention, consists of 15 characters arranged into four fragments or elements; the elements are separated by an underscore, a hyphen or a period. Some SCADA Human Machine Interface (HMI) software may not allow the use of some of these delimiters, so the user will have to use whatever works best in their system.

The complete tagname follows a standard format, providing successive detail about the signal, its source and the specifics of the signals, in each fragment or element. Each field has defined alphanumeric codes or phrases which can be organized into 'dictionaries'; to construct a tagname, one assembles phrases from the various dictionaries to build a successively more detailed definition of the signal. This TSNC applies to both hardware or field signals as well as to the internal software or program points. Characters in the fourth field identify the nature of the signal and whether it is a field or a virtual type signal.

For example, a pump at the New Age Pumping Station would be identified with the initial fragments of NAP_PP05. The discharge valve for the pump would be tagged as NAP_PP05_DV. And the open status limit switch on the valve would complete the tagname as NAP_PP05_DV_SOD. The last fragment, SOD, is composed of a signal type, S, and a status field, OD.

B.1.1 Structure and Format of Tagnames

The general format of the TSNC tagnames is as follows, using the underscore to separate the fragments of the tagname; alternatively, the hyphen or period could also be used to separate the fragments of the tagname.

LLL_ EEEE_ CC_ TSS

where

LLL is the Location within the overall system,
EEEE is the major Equipment,

CC	is the Component of the equipment,
T	is the Type of Signal, including field or virtual,
SS	is the Signal Designation.

Note that the fourth fragment consists of two parts: a signal type (**T**) and a signal designation (**SS**); hence, this fragment serves to identify both the type of information as well as the signal detail.

The **LLL** element identifies the geographic location of the equipment. These three alphanumeric characters define the physical location in the overall SCADA system. This fragment may be further divided into two characters for the location and one character for the type of facility. For example, HLF could denote the High Lift Station of a water treatment plant; NA indicates New Age station and the P denotes the pumping station; NAR indicates New Age and the R denotes the reservoir.

The **EEEE** element identifies the actual equipment or device, including the unit number. This fragment can be subdivided with some characters for the equipment, and some characters for the number. Two letters for equipment allows for two digits for the unit number; three letters for the equipment results in one letter for the unit; four letters could be used to designate the complete facility. Consider the following examples:

HLPx	High Lift Station Pump x
BPVx	Bypass Valve x
BLDG	Building (no number)
FLxx	Filter xx, range 1–99
PPxx	Pump xx, range 1–99

The **CC** element identifies the component of subsystem of the main equipment or device. If the entire equipment is being referenced as a single unit, then the CC field could be '00'. Typically, the CC field indicates the part of the equipment being referenced such as DV for discharge valve, IV for inlet valve and DG for drain gate. Some examples of component fragments are shown below:

RWP2_DV	Raw Water Pump 2, Discharge Valve
RWL1_FT	Raw Water Line 1, Flow Transmitter
TWP4_PT	Treated Water Pump 4, Pressure Transmitter

One additional use for the component element is to designate one of several unit signals associated with the basic equipment element. For example, a motor may have four RTD temperature sensors associated with it, so the component field can be the numbered sensor:

HLF_HLP5_01_ATA	pump #5 temperature sensor #1
HLF_HLP5_02_ATA	pump #5 temperature sensor #2
HLF_HLP5_03_ATA	pump #5 temperature sensor #3
HLF_HLP5_04_ATA	pump #5 temperature sensor #4

The **TSS** element combines two parts: the signal type, T, and the signal designation, SS. Since one of the goals of the TSNC is to adopt a universal naming system for both the field hardware signals and the internal program points, the signal type, T, indicates which of several point types applies to the tagname.

The **T** element, or Signal Type, will be one of the following:

S	discrete status input	X	virtual discrete/boolean input point
A	discrete alarm status input	Y	virtual discrete/boolean output point
D	discrete on/off output	Z	virtual boolean flag point
Q	analog value input	N	virtual analog input; internal value
K	analog value output	R	virtual analog output; internal value

Note that the virtual points are those for which there is no hardware or field counterpart; it is not wired to anything in the field. For example, if the two analog input signals, HLF_RWL1_FT_QFI and HLF_RWL2_FT_QFI are added together, the sum may be stored in a virtual total, HLF_RWL0_FI_NFI.

The **SS** element, or Signal Designation, identifies the specific signal associated with the specified equipment. This fragment relies heavily on the ISA naming convention. While this ISA system does identify the nature of the signal, it does not identify the location or the type of equipment being referred to. The ISA tagging system was illustrated in Section 5.1 previously.

Some examples of signal designation fragments are shown below:

RWP3_SS_SRC	Remote Control Selector
MIV2_EV_DON	Open Control on valve
AST1_LS_ALH	Alum Tank High Level Sensor
BSP2_PF_ZRM	Pump 2 Remote Manual Selected

B.1.2 Tagname Fragment Dictionaries

As stated earlier, the TSNC uses a four fragment system in which each fragment identifies one part of the complete tagname. A series of 'dictionaries' may prove useful in constructing tagnames for a SCADA system. As the system is designed and developed, additional fragments may be added to these dictionaries. At the end of the project, there will be a complete set of listings for all fragments of the tagnames used in the SCADA project.

Each of the four fragments or elements of the tagname will have a list of assigned phrases or definitions. The following subsections contain a series of dictionaries for the tagname fragments, based upon the water treatment plant application. Of course, this system may be modified and tailored in any way desired; this is meant as a design methodology and provide a starting point for tagname construction.

B.1.3 Fragment #1 Location Fragments Listings

CHM	Chemical Systems
FLT	Filtration Facilities
HLF	High Lift Station
LLF	Low Lift Station
MCC	Motor Control Centre
GRM	Generator Room
PSA	Pumping Station Acme
PSN	Pumping Station New Age

PSW	Pumping Station Widget
PRE	Pretreatment Facility
RMF	Residue Management Facility
WTP	Water Treatment Plant

B.1.4 Fragment #2 Equipment Fragments Listings

'A'	ACP	Activated Carbon Metering Pump
	ACT	Activated Carbon Storage Tank
	AFD	Adjustable Frequency Drive
	AFP	Alum Feed Pump
	ALP	Alum Metering Pump
	AST	Alum Storage Tank
	ATP	Alum Transfer Pump
'B'	BLG/BLDG	Building (general)
	BPP	Bypass Pump
	BPV	Bypass Valve
	BSP	Booster Station Pump
'C'	CCC	Chlorine Contact Tank
	CFP	Carbon Feed Pump
	CLA	Clarifier
	CFP	Chlorine Feed Pump
	CHT	Chlorine Hypochlorite Tank
	CGD	Chlorine Gas Detector
	CLC	Chlorinator
	CMP	Chemical Metering Pump
	CWL	Clearwell
'D'	DFP	Diesel Fuel Pump
	DFT	Diesel Fuel Tank
	DIAL	Alarms Autodialer
	DIH	Discharge Header
'E'	EFH	Effluent Header
	ELR	Electrical Room
	EPG	Emergency Power Generator
	EQP	Equipment (Failure)
'F'	FBH	Filter Bypass Header
	FIC	Filter Inlet Chamber
	FTP	Fuel Transfer Pump
	FLC	Flocculator
	FLT	Flocculation Tank
	FMX	Flash Mixer
	FUP	Fluoride Feed Pump
	FIL	Filter
'G'	GDS	Gas Detection System
	GEN	Generator (electric)
	GWP	Ground Water Pump

'H'	HLP	High Lift Pump
	HFS	HFS Feed Pump
	HLW	High Lift Well
	HTK	Holding Tank
'I'	ILP	Indicator Lamp
	INL	Inlet Line
	INC	Inlet Chamber
'L'	LPN	Lighting Panel
	LLP	Low Lift Pump
	LLS	Low Lift Station
	LLW	Low Lift Well
'M'	MCC	Motor Control Centre
	MCS	Microstrainer
	MIV	Motorized Isolation Valve
	MXR	Mixer (general)
'P'	PFP	Polymer Feed Pump
	PFS	Polymer Feed System
	PLC	Programmable Logic Controller
	PLNT	Plant (general)
	PP	Pump (general)
	PSD	Pumping Station Discharge
	PSI	Pumping Station Inlet
	PTP	Polymer Transfer Pump
'R'	RES	Reservoir Cell
	RWI	Raw Water Intake
	RWL	Raw Water Line
'S'	SAP	Sample Pump
	SCN	Bar Screen
	SHP	Sodium Hypochlorite Pump
	SIP	Silica Metering Pump
	SIT	Silica Storage Tank
'T'	TFS	Transfer Switch
	TSC	Travelling Screen
	TWL	Treated Water Line
	TXR	Transformer
'U'	UHT	Heater Unit
	UPS	Uninterruptible Power Supply
	UVS	Ultra Violet (UV) System
'V'	VCR	Valve Chamber
	VV	Valve – General
'W'	WEL	Well
	WEP	Well Pump
	WST	Wastewater Storage Tank
	WWL	Wet Well
	WWP	Wastewater Pump
	WWT	Wastewater Tank

B.1.5 Fragment #3 Component Fragments Listings

Analyzers	AN	Analyzer Unit
	RC	Residual Chlorine
Circuit Breaker	FB	Feeder Breaker
	MB	Main Breaker
	PL	Peak Load
Controllers	FC	Flow Controller
	LC	Level Controller
	SC	Stroke Controller
	UC	Unit (Packaged) Controller
Drives	GD	Gear Drive
	VFD	Variable Frequency Drive
	VSD	Variable Speed Drive
Gates	AG	Alternate Gate
	BG	Bypass Gate
	DG	Drain Gate
	FG	Flood Gate
	IG	Inlet Gate
	OG	Outlet Gate
	SG	Sluice Gate
	WG	Weir Gate
Motors	DM	Drive Motor
	MM	Modulating Motor
Sensors	AE	Analyzer Element
	FE	Flow Element
	GE	Gas Sensing Element
	IE	Current Element
	LE	Level Element
	PE	Pressure Element
	SE	Speed Element
	TE	Temperature Element
	VE	Vibration Element
	ZE	Position Element
Switches	DS	Door Switch
	FS	Flow Switch
	HS	HOA Switch
	KS	Key Switch
	LS	Level Switch
	PS	Pressure Switch
	SS	Selector Switch
	TS	Temperature Switch
	ZS	Limit Switch
Transmitters	FI	Flow Indication
	FT	Flow Transmitter
	LI	Level Indication

	LT	Level Transmitter
	PI	Pressure Indication
	PT	Pressure Transmitter
	TI	Temperature Indication
	TT	Temperature Transmitter
	ZT	Position Transmitter
Valves	AV	Agitator Valve
	BV	Bypass Valve
	CV	Check Valve
	DV	Discharge Valve
	EV	Effluent Valve
	FV	Flow Valve
	GV	Isolating Valve
	MV	Modulating (Control) Valve
	NV	Drain Valve
	OV	Outlet Valve
	PV	Pressure Relief Valve
	SV	Solenoid Valve
	WV	Surface Wash Valve
	ZV	Backwash Valve
Units (general)	01	Unit 1
	02	Unit 2, and so on
	PP	Program Point (internal virtual)
Miscellaneous	CF	Configuration
	CP	Control Panel
	EP	Emergency Power
	HN	Horn (Annunciator)
	MX	Mixer
	PB	Pushbutton
	PF	Program Flag
	PL	Pilot Lamp
	RT	Runtime
	TU	Turbidity

B.1.6 Fragment #4a Signal Type Fragments Listings

Field I/O	S	Status (Digital) Input
	A	Alarm (Digital) Input
	D	Control (Digital) Output
	Q	Quantity (Analog) Input
	K	Control/Setpoint (Analog) Output
Virtual I/O	X	Digital Input
	Y	Digital Output
	Z	Digital Value/Flag
	N	Analog Input
	R	Analog Output
	V	Analog Value

B.1.7 Fragment #4b Designation Fragments Listings

Status Inputs

BM	Bypass Mode
CD	Closed Indication
CM	Control Mode
DS	Duty Select
DN	Done/Complete
HI	High Level
LC	Local Control Mode
MC	Manual/Local Control Mode
OD	Open Indication
OF	Off Status
ON	On Status
PF	Program Flag
RA	Remote Automatic Mode
RC	Remote Control
RD	Ready Indication
RM	Remote Manual Mode
RN	Running
RS	Reset
SE	Select
SS	Status (General)
SP	Stop
SR	Stop Required
ST	Start

Alarm Inputs

	1st Character		*2nd Character*
A	Gas	A	Alarm
B	Pre	B	Bus
C	Analytical	D	Differential
D	Density	E	Illegal
E	Voltage	F	Fault
F	Flow	H	High
G	General	I	Low Low
H	Torque	L	Low
I	Current	M	High High
J	Power	N	Input
K	Time	O	Output
L	Level	R	Failure
N	Ground	W	Warning
O	Oxygen		
P	Pressure		
R	Overload		
S	Entry/Exit		
T	Temperature		
V	Vacuum		
W	Weight		

Control Outputs

AI	Alarm Indication
AU	Automatic Mode

	CE	Close
	DE	Decrease (Down)
	DN	Done/Complete
	FI	Failure Indication
	FW	Forward
	HI	High Speed
	IN	Increase (Up)
	IL	Interlock
	LC	Local Control
	LO	Low Speed
	ON	Open
	RA	Remote Automatic Mode
	RC	Remote Control Mode
	RM	Remote Manual Mode
	RN	Run
	RS	Reset
	RV	Reverse
	SE	Select
	SP	Stop
	ST	Start
Quantity Inputs	CI	Conductivity
	DI	Density
	EI	Voltage
	FI	Flow
	HZ	Frequency
	HI	pH
	II	Current
	JI	Power
	LD	Level Differential
	LI	Level
	NI	Turbidity
	OI	Dissolved Oxygen
	PD	Pressure Differential
	PI	Pressure
	PV	Program Variable
	RC	Residual Chlorine
	RF	Residual Fluoride
	SC	Streaming Current
	SI	Speed Indication
	SR	Status Register
	TI	Temperature
	WI	Weight
	ZI	Position
Setpoint Outputs	SP	Speed Setpoint
	ZP	Position Setpoint

Appendix C: Sample Process Control Logic Description

Appendix C contains a complete Process Control Logic Description (PCLD) as described in detail in chapter 7.

The following PCLD is for the High Lift Pumping Station at a water treatment plant; this facility discharges treated water out into the water distribution system. This PCLD is based upon a project completed for a water treatment plant and distribution system; the information follows the format as described in Chapter 7.

High Lift Station

The high lift station within the water treatment plant receives treated water from the filtration system of the plant. The pumps are started and stopped so as to maintain the level in an elevated water tank between setpoint limits.

C.1 System Control Strategy Overview

The high lift pumps are operated to draw potable water from the underground reservoir within the water treatment plant, also referred to as the High Lift Clearwell, and pump the water into the water distribution system.

C.1.1 Overview of Process Area

The process involves pumping water from an underground reservoir within the water plant out into the distribution system. The Harrow Elevated Tank, located some 10 km away, serves as a source of water and as a pressure-maintaining facility. The level signal for the tank is used as the primary control parameter for determining when the Programmable Process Controller (PPC) should start and stop the high lift pumps. In the event of the failure of this signal, typically due to signal loss over the telephone line, the plant discharge pressure is used as the secondary control parameter.

C.1.2 Summary of Operational Concepts

Each of the three high lift pumps has an associated operating range of levels for the elevated tank; as the tank level drops below the start level setpoint for one pump, that pump is started and the previously running pump is stopped; as the tank level drops below the start level setpoint for the second pump, the second pump is started and the previous pump is stopped.

Similarly, as the tank level rises, the current pump is stopped and the previous pump is started; this control strategy ensures that only one high lift pump is running at any time.

When the level signal is not available, then high lift pump 2 is operated based upon the plant discharge pressure and the start and stop pressure setpoints.

C.2 Facilities and Parameters

The high lift pumping station equipment, major signals and parameters are described below.

C.2.1 Summary of Equipment

The high lift station consists of three fixed speed pumps which are assigned duties based upon their respective start and stop level setpoints. A below grade reservoir, the high lift clearwell, receives treated water from the filtration system of the plant. Water is pumped from this clearwell out into the distribution system. The discharge flowrate and pressure are used in the operation of the pumps.

C.2.2 Key Process Signals

The following signals are continuously monitored by the high lift PPC and are used in the control of the high lift pumps:

- High Lift Clearwell Level
- Elevated Water Tank Level
- High Lift Discharge Pressure
- Ground Water Level

C.2.3 Control Parameters and Setpoints

Control setpoints are entered by the operator through the SCADA user workstations; the setpoint parameters used by the high lift PPC control program are as follows:

- Water Tower Pump 1 Start Level
- Water Tower Pump 2 Start Level
- Water Tower Pump 3 Start Level
- Water Tower Pumps Stop Level

- Plant Discharge Pump 2 Start Pressure
- Plant Discharge Pump 2 Stop Pressure
- Water Tower Maximum Level
- Reservoir Minimum Level
- High Lift Clearwell Minimum Level
- High Lift Pump Start Delay Time
- Reservoir Ground Level Differential

C.3 Control Logic Description

When the tower level drops below the Pump 1 Start Level setpoint, then pump 1 is started. If the tower level drops below the Pump 2 Start Level setpoint, then pump 2 is started and pump 1 is stopped. If the tower level drops below the Pump 3 Start Level setpoint, then pump 3 is started and pump 2 is stopped.

When the tower level rises above the Pump 3 Start Level setpoint, then pump 3 is stopped and pump 2 is started. When the tower level rises above the Pump 2 Start Level setpoint, then pump 2 is stopped and pump 1 is started. When the tower level rises above the Pump 1 Start Level setpoint, then pump 1 is stopped.

The duty assignments of the three high lift pumps to the duty numbers 1, 2, 3 are made by the operator through the operator workstation. Any pump can be assigned to any duty.

In the event that the tower level signal is continuously lost for 15 min, the control program will automatically switch control action to start and stop the high lift pump 2 based upon the plant effluent discharge pressure. The program will remain in this mode until the level signal is restored.

For the discharge pressure control mode, when the discharge pressure drops below the Pump 2 Start Pressure setpoint, then the high lift pump 2 is started. If the discharge pressure rises above the Pump 2 Stop Pressure setpoint, then the high lift pump 2 is stopped. Note that only one high lift pump is used during the discharge pressure mode of control.

The control program continuously compares the Reservoir Level with the Ground Water Level and compares the difference with the Reservoir Ground Water Level Difference setpoint. If the difference is less than the setpoint, then a warning is generated to the operator through the SCADA system.

The high lift pumps can be operated in any of three (3) modes: Local, Remote Manual and Remote Automatic; each of these is described in detail following.

- Local – Manual starting and stopping of the high lift pump is performed using the local HOA switch located on the control panel for the pump;
- Remote Manual – Pump is started and stopped by commands received from the host SCADA system; the operator clicks on 'Start' and 'Stop' pushbuttons on process graphic displays;
- Remote Automatic – Pump is started and stopped based upon the start and stop tower level (or discharge pressure) setpoints for the pump, as described previously.

C.3.1 General Requirements

There are some checks made by the control program before any pump is enabled to run under automatic control:

- Reservoir Level must be above the Reservoir Minimum Level setpoint;
- Water Tower Level must be below the Water Tower Maximum Level setpoint.

With the high lift pumps enabled based upon satisfying the above conditions, the control program operates three high lift pumps based upon their start and stop level setpoints, as entered by the operator. The high lift pumps are started and stopped to maintain the level within the elevated tower.

There is an internal software flag maintained by the Programmable Logic Controller (PLC) control program to indicate which of the two control variables the program is using for control: elevated tank level or plant discharge pressure. When the tower level signal is lost, then the program sets the control mode flag to indicate pressure control. When the tower level signal is restored and remains valid for 3 min, then the program clears the control mode flag to indicate level control and reverts back to the normal level control.

The control program maintains runtimes of each of the pumps in hours and tenths of hours. These values can be reset by the operator via the SCADA system.

C.3.2 Local Control Mode

The Hand-Off-Auto (HOA) switch on the local control panel for the high lift pump must be in the 'Hand' position to run the pump and in the 'Off' position to stop the pump. The PLC control program has no control in this mode of operation.

C.3.3 Remote Manual Control Mode

The HOA switch for the pump must be in the 'Auto' position, which allows the PLC control program to start and stop the pump, based upon commands received from the host SCADA system. An internal Manual/Automatic mode flag is maintained by the program to determine which remote mode is in effect. From the SCADA process graphic display, the operator can select the 'Remote Manual' mode of control.

The operator can click on 'Start' and 'Stop' pushbuttons on a process graphic display to issue the start/stop commands to the PLC control program. This causes the control program to run or stop the corresponding high lift pump.

The control program will override the checks described in Section C4.1 so that the operator can control any of the pumps directly from the SCADA system.

C.3.4 Remote Automatic Control Mode

The HOA switch for the pump must be in the 'Auto' position, which allows the PLC control program to start and stop the pump, based upon the start and stop tower levels assigned to the duty pump through the host SCADA system. The internal

Manual/Automatic mode flag maintained by the program must be set to 'Remote Automatic' mode of control by the operator through the graphic display.

The control program will start and stop the high lift pumps based upon their respective start and stop level setpoints. If any of the permissive conditions described above are not met, then the high lift pumps will be prevented from starting.

When the tower level drops below a pump start level setpoint, then the control program will start the corresponding pump and stop the previously running pump. When the tower level rises above the pump stop level setpoint, then the control program will stop the corresponding pump and start the previous pump.

In the discharge pressure mode of control, the single pump is started and stopped in the same manner, except that the plant effluent pressure is used in place of the tower level for the control variable, and the pressure start and stop setpoints are used in place of the tower start and stop setpoints.

C.4 Special Considerations

There are some checks and interlocks incorporated into the system to prevent improper operation of the high lift pumps; each is described below, along with the handling of failures of equipment and alarm conditions.

C.4.1 Software Interlocks

The following conditions in the program will prevent a high lift pump from starting, if the pump is operating in the Remote Automatic mode; if any of these conditions is True, an alarm is generated to the SCADA system:

- The Reservoir Level is less than the Reservoir Minimum Level setpoint;
- The Elevated Tank Level is greater than the Tower Maximum Level setpoint.

C.4.2 Hardwired Interlocks

There are no hardwired interlocks associated with the high lift pumps.

C.4.3 Failures and Alarms

The control program maintains software and hardware failure flags for the high lift pumps. If a failure of a pump occurs, then the virtual (software) flag will be set by the program, and an alarm will be generated through the SCADA system to notify the operator.

If a pump does not start within the time specified by the High Lift Pump Start Delay Time setpoint, then a 'Pump Start Failure' alarm is generated to SCADA. The pump is flagged as failed, and the backup duty pump will be operated in its place.

If a pump's HOA switch is not in the 'Auto' position for more than 15 min, then a 'Pump Not In Auto' alarm is generated to SCADA.

If a pump runs for more than 8 h continuously, then a 'Pump Runtime Exceeded' alarm is generated to SCADA.

Any of these software interlock alarms and failure alarm conditions will cause the program to set the pump's virtual alarm/failure flag.

If the difference in levels between the Ground Water and the Reservoir is less than the difference setpoint, as described earlier, then a warning is generated through the SCADA system.

C.4.4 Software Interfacing

There are two PPCs for this water treatment plant system: Filter PPC which handles all operations associated with the two filters; and the Plant PPC which handles the Low Lift, Clarifier, High Lift and Chemical systems. There is a semaphore signal between the two PPCs, but this is addressed in another PCLD.

Appendix D: Program Listings for PPC Program

Appendix D provides the complete PPC program listings for the sample project described in chapter 12. The purpose of this appendix is to illustrate a real world example of a complete PPC application program.

The example program included in this appendix is intended to illustrate the suggested structured programming techniques as outlined in this book. The application of tagnames, as shown in this book, is used to illustrate the use of the Tagname Signal Naming Convention system for naming both field I/O points and internal software program points. The program is organized into subroutines to illustrate the structured approach to programming for Programmable Logic Controllers (PLCs). A brief description of the design follows.

A water reservoir or tank includes a level transmitter whose level signal is used for the automatic control of the two pumps. The logic is designed to start and stop the pumps based upon high and low levels in the reservoir when operating in automatic mode; in manual mode, the pumps can be controlled using the field input pushbuttons on the PLC demonstration unit. Selecting the mode of operation is accomplished with virtual tags in the database that relate to a Human–Machine Interface (HMI) application; this HMI facility is not included in this program.

Alarm conditions are checked for pump start/stop and discharge valve open/close operation and high, high high, low and low low levels in the reservoir. A failure to start or stop within an allowed time, or a failure for the discharge valve to open or close within an allowed time, results in the setting of a virtual software alarm flag.

Pump control mode must be in the Remote mode, as indicated by a field selector input; in this mode, the pump can be selected for either Remote Manual or Remote Automatic mode of control. The selection of remote control modes is achieved through software points originating from an HMI application. The starting and stopping of the pump, and the opening and closing of the discharge valve, are handled by a pump control subroutine.

D.1 Program Structure

This project consists of one continuous main program with subroutines and one periodic or interval task and program with one program. Each is described below.

The Main Continuous Program is organized as shown below.

Main Routine	Mainline routine for program; calls other subroutines
General Functions	General operations associated with entire application; real time clock of processor is accessed for time

Process Inputs	Handle discrete and analog input signals
Alarms	Checks for pump start/stop and discharge valves operating within allowed time; high, high high, low and low low alarm conditions for the reservoir level
Automatic Control	Logic for selecting the current lead duty pump; operating the lead duty pump in automatic mode; generating alarm levels for reservoir
Pump Control	Contains logic for selecting the mode of control (Manual or Automatic), starting and stopping the pump in either mode and opening and closing the discharge valve with the pump
Process Outputs	Handle discrete and analog output signals based upon logic
Effluent Valve	Logic to operate the effluent valve in Remote Manual mode

The Synchronous Task consists of a single program which executes at fixed intervals. The program consists of the following logic:

Data Collection	Accumulates runtime for the two pumps; cascading timers are used

The complete program listings for the pumping station controller are as follows.

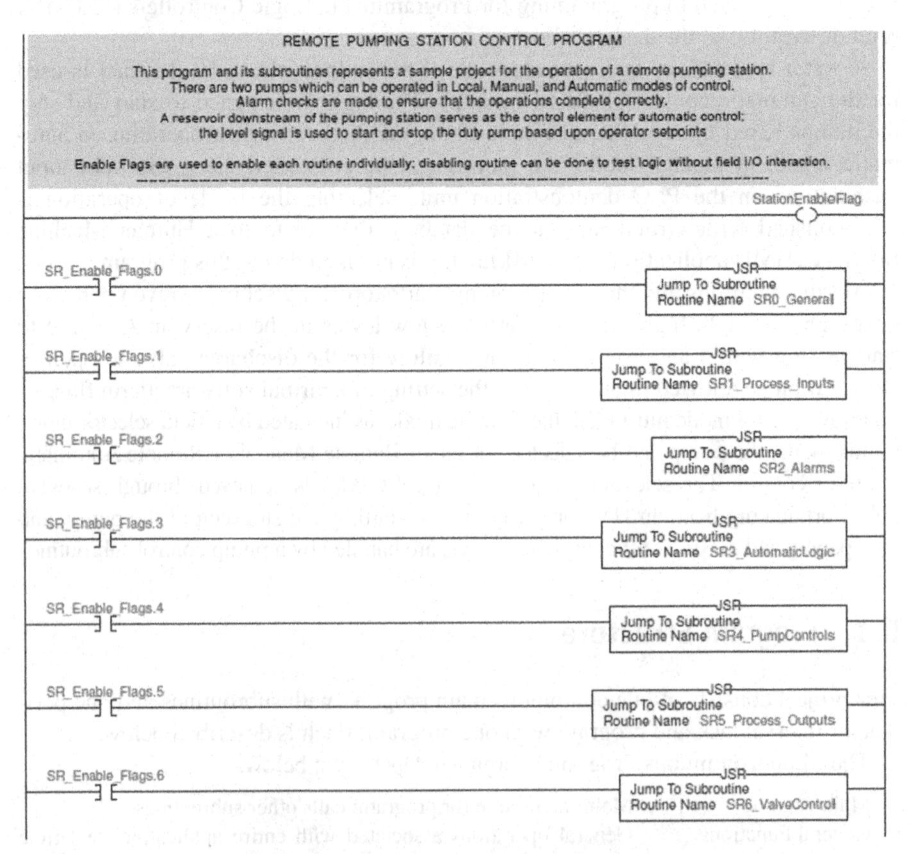

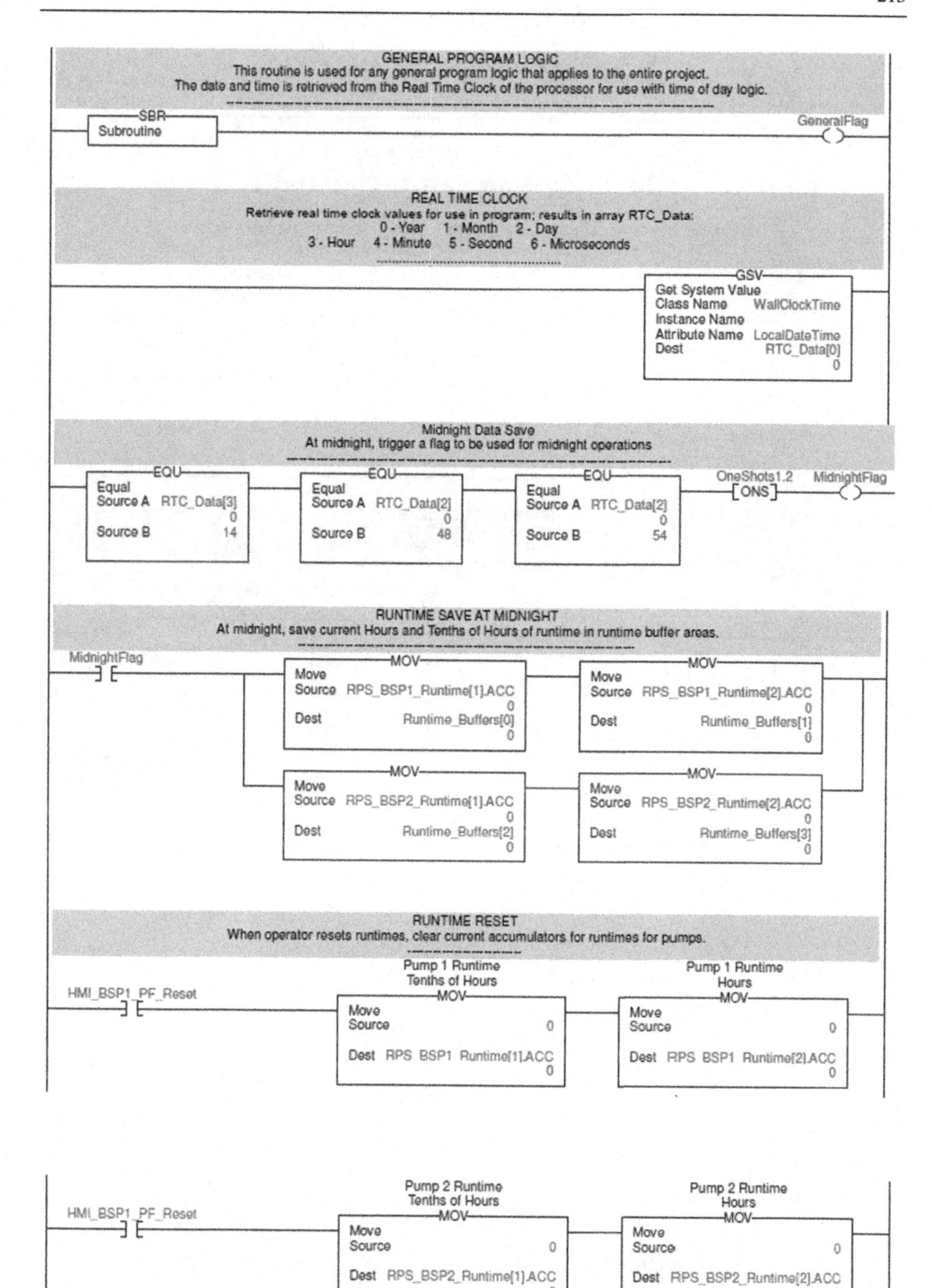
GENERAL PROGRAM LOGIC
This routine is used for any general program logic that applies to the entire project.
The date and time is retrieved from the Real Time Clock of the processor for use with time of day logic.
SBR
Subroutine
GeneralFlag
REAL TIME CLOCK
Retrieve real time clock values for use in program; results in array RTC_Data:
0 - Year 1 - Month 2 - Day
3 - Hour 4 - Minute 5 - Second 6 - Microseconds
GSV
Get System Value
Class Name WallClockTime
Instance Name
Attribute Name LocalDateTime
Dest RTC_Data[0]
0
Midnight Data Save
At midnight, trigger a flag to be used for midnight operations
EQU
Equal
Source A RTC_Data[3]
0
Source B 14
EQU
Equal
Source A RTC_Data[2]
0
Source B 48
EQU
Equal
Source A RTC_Data[2]
0
Source B 54
OneShots1.2
ONS
MidnightFlag
RUNTIME SAVE AT MIDNIGHT
At midnight, save current Hours and Tenths of Hours of runtime in runtime buffer areas.
MidnightFlag
MOV
Move
Source RPS_BSP1_Runtime[1].ACC
0
Dest Runtime_Buffers[0]
0
MOV
Move
Source RPS_BSP1_Runtime[2].ACC
0
Dest Runtime_Buffers[1]
0
MOV
Move
Source RPS_BSP2_Runtime[1].ACC
0
Dest Runtime_Buffers[2]
0
MOV
Move
Source RPS_BSP2_Runtime[2].ACC
0
Dest Runtime_Buffers[3]
0
RUNTIME RESET
When operator resets runtimes, clear current accumulators for runtimes for pumps.
HMI_BSP1_PF_Reset
Pump 1 Runtime
Tenths of Hours
MOV
Move
Source 0
Dest RPS_BSP1_Runtime[1].ACC
0
Pump 1 Runtime
Hours
MOV
Move
Source 0
Dest RPS_BSP1_Runtime[2].ACC
0
HMI_BSP1_PF_Reset
Pump 2 Runtime
Tenths of Hours
MOV
Move
Source 0
Dest RPS_BSP2_Runtime[1].ACC
0
Pump 2 Runtime
Hours
MOV
Move
Source 0
Dest RPS_BSP2_Runtime[2].ACC
0
RET
Return from Subroutine

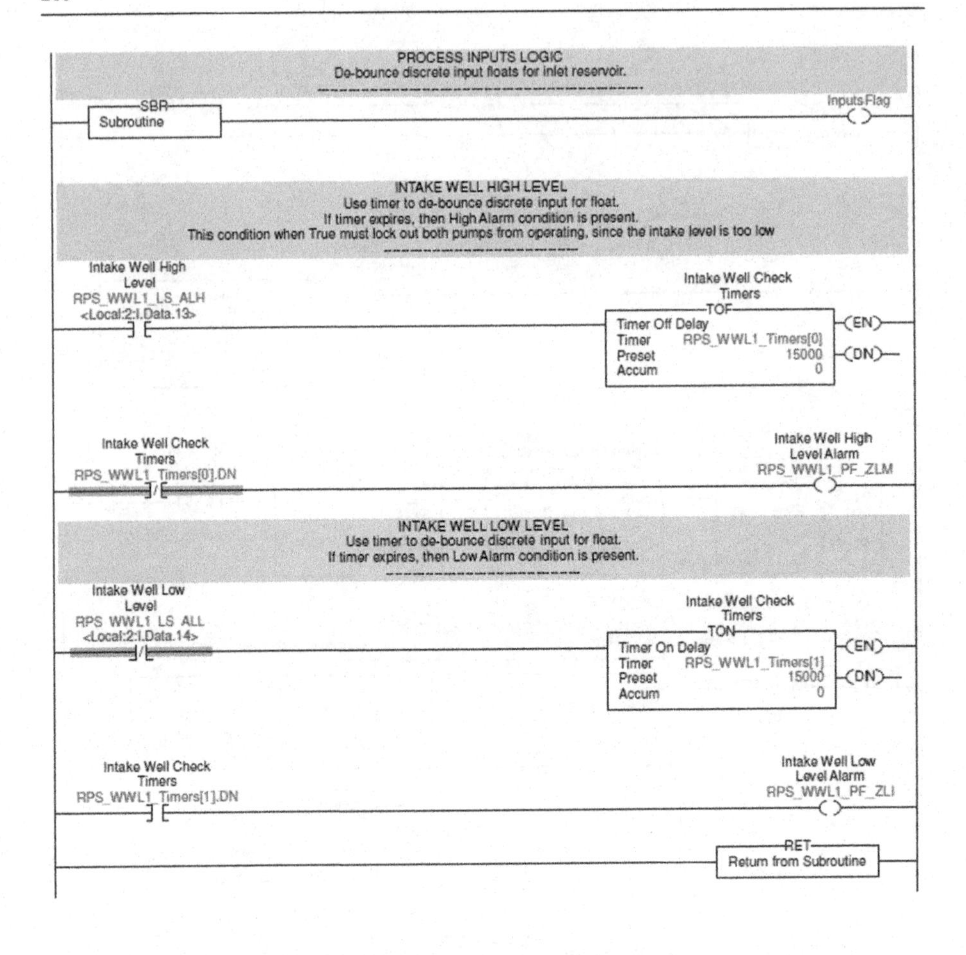

PROCESS INPUTS LOGIC
De-bounce discrete input floats for inlet reservoir.
SBR
Subroutine
Inputs Flag
INTAKE WELL HIGH LEVEL
Use timer to de-bounce discrete input for float.
If timer expires, then High Alarm condition is present.
This condition when True must lock out both pumps from operating, since the intake level is too low
Intake Well High Level
RPS_WWL1_LS_ALH
<Local:2:I.Data.13>
Intake Well Check Timers
TOF
Timer Off Delay
Timer RPS_WWL1_Timers[0]
Preset 15000
Accum 0
EN
DN
Intake Well Check Timers
RPS_WWL1_Timers[0].DN
Intake Well High Level Alarm
RPS_WWL1_PF_ZLM
INTAKE WELL LOW LEVEL
Use timer to de-bounce discrete input for float.
If timer expires, then Low Alarm condition is present.
Intake Well Low Level
RPS_WWL1_LS_ALL
<Local:2:I.Data.14>
Intake Well Check Timers
TON
Timer On Delay
Timer RPS_WWL1_Timers[1]
Preset 15000
Accum 0
EN
DN
Intake Well Check Timers
RPS_WWL1_Timers[1].DN
Intake Well Low Level Alarm
RPS_WWL1_PF_ZLI
RET
Return from Subroutine

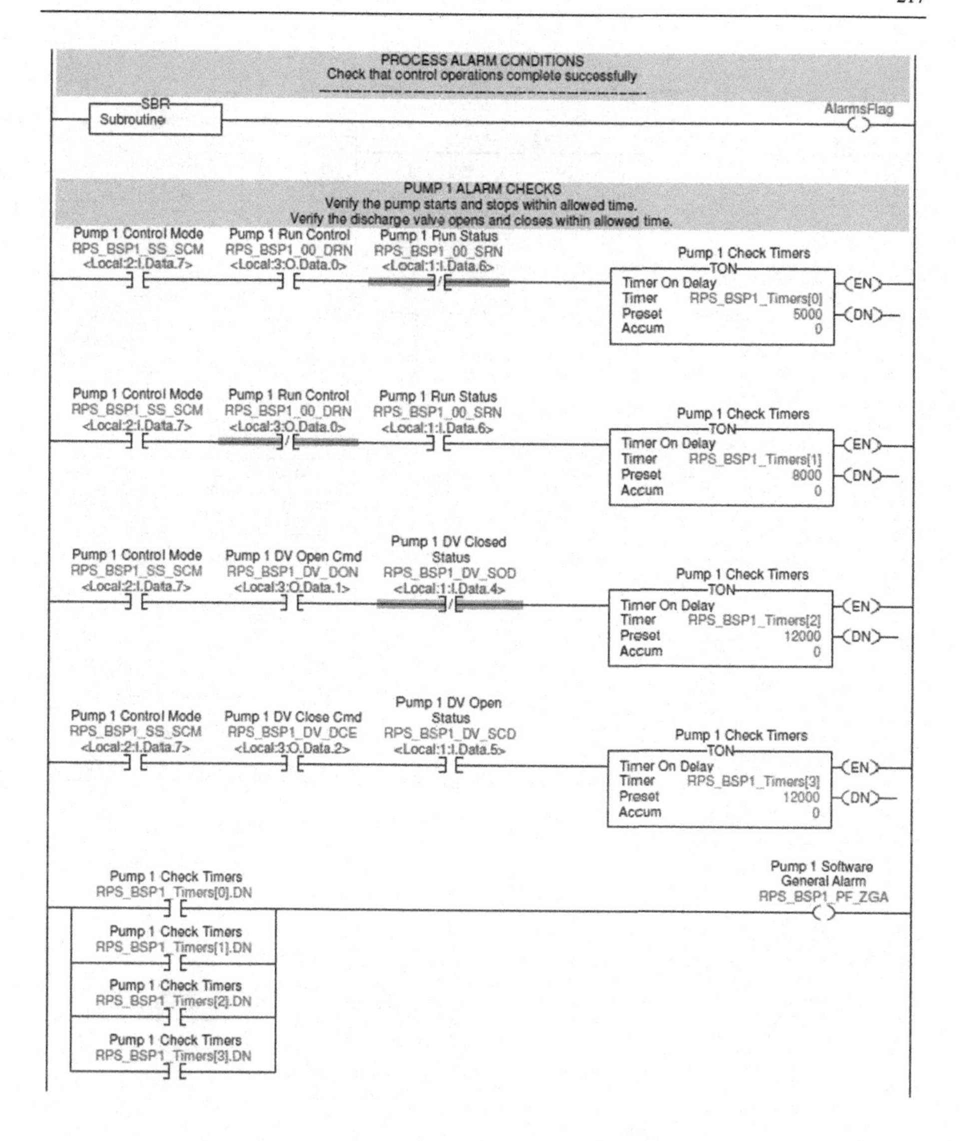

PROCESS ALARM CONDITIONS
Check that control operations complete successfully
SBR
Subroutine
AlarmsFlag
PUMP 1 ALARM CHECKS
Verify the pump starts and stops within allowed time.
Verify the discharge valve opens and closes within allowed time.
Pump 1 Control Mode
RPS_BSP1_SS_SCM
<Local:2:I.Data.7>
Pump 1 Run Control
RPS_BSP1_00_DRN
<Local:3:O.Data.0>
Pump 1 Run Status
RPS_BSP1_00_SRN
<Local:1:I.Data.6>
Pump 1 Check Timers
TON
Timer On Delay
Timer RPS_BSP1_Timers[0]
Preset 5000
Accum 0
EN
DN
Pump 1 Check Timers
Timer RPS_BSP1_Timers[1]
Preset 8000
Pump 1 DV Open Cmd
RPS_BSP1_DV_DON
<Local:3:O.Data.1>
Pump 1 DV Closed Status
RPS_BSP1_DV_SOD
<Local:1:I.Data.4>
Timer RPS_BSP1_Timers[2]
Preset 12000
Pump 1 DV Close Cmd
RPS_BSP1_DV_DCE
<Local:3:O.Data.2>
Pump 1 DV Open Status
RPS_BSP1_DV_SCD
<Local:1:I.Data.5>
Timer RPS_BSP1_Timers[3]
Preset 12000
RPS_BSP1_Timers[0].DN
RPS_BSP1_Timers[1].DN
RPS_BSP1_Timers[2].DN
RPS_BSP1_Timers[3].DN
Pump 1 Software General Alarm
RPS_BSP1_PF_ZGA

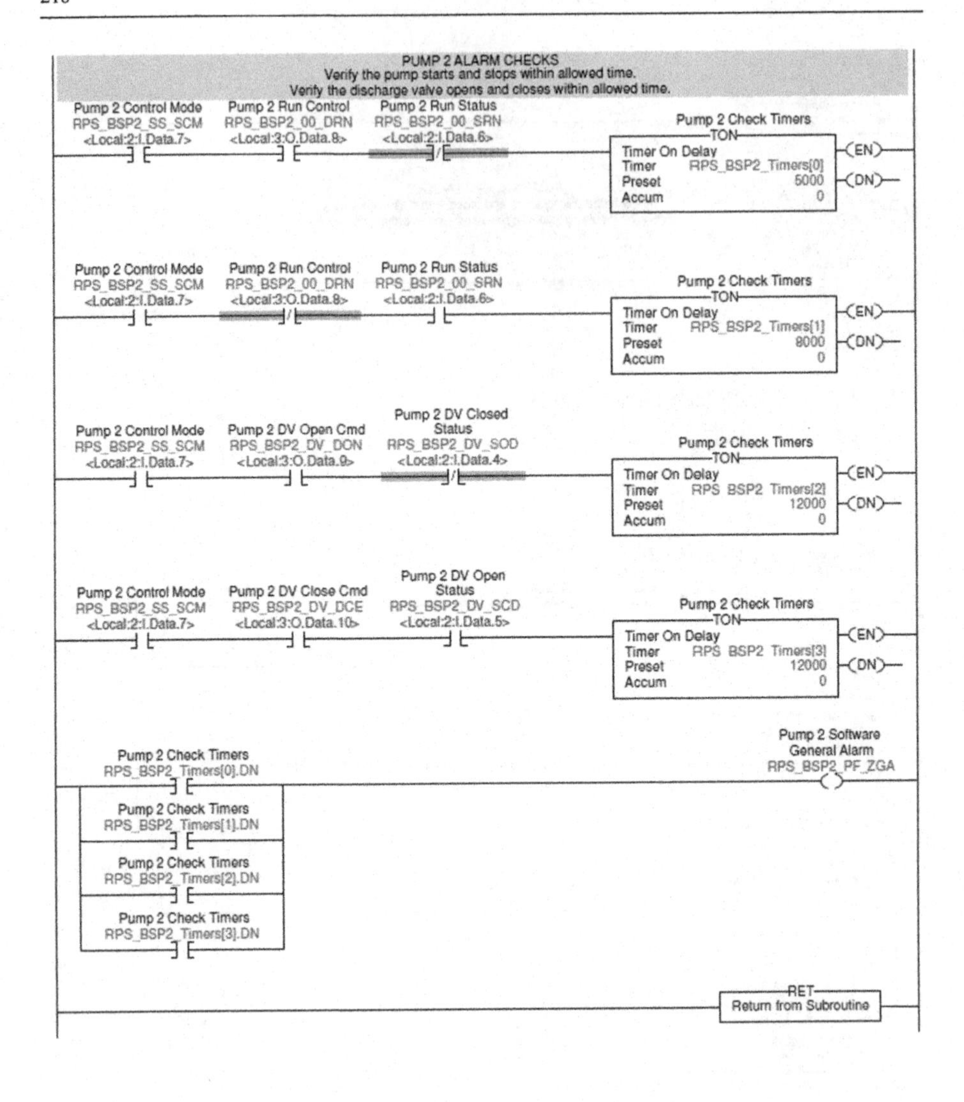
PUMP 2 ALARM CHECKS
Verify the pump starts and stops within allowed time.
Verify the discharge valve opens and closes within allowed time.
Pump 2 Control Mode
RPS_BSP2_SS_SCM
<Local:2:I.Data.7>
Pump 2 Run Control
RPS_BSP2_00_DRN
<Local:3:O.Data.8>
Pump 2 Run Status
RPS_BSP2_00_SRN
<Local:2:I.Data.6>
Pump 2 Check Timers
TON
Timer On Delay
Timer RPS_BSP2_Timers[0]
Preset 5000
Accum 0
EN
DN
Pump 2 Check Timers
Timer RPS_BSP2_Timers[1]
Preset 8000
Accum 0
Pump 2 DV Open Cmd
RPS_BSP2_DV_DON
<Local:3:O.Data.9>
Pump 2 DV Closed Status
RPS_BSP2_DV_SOD
<Local:2:I.Data.4>
Timer RPS_BSP2_Timers[2]
Preset 12000
Accum 0
Pump 2 DV Close Cmd
RPS_BSP2_DV_DCE
<Local:3:O.Data.10>
Pump 2 DV Open Status
RPS_BSP2_DV_SCD
<Local:2:I.Data.5>
Timer RPS_BSP2_Timers[3]
Preset 12000
Accum 0
Pump 2 Check Timers
RPS_BSP2_Timers[0].DN
RPS_BSP2_Timers[1].DN
RPS_BSP2_Timers[2].DN
RPS_BSP2_Timers[3].DN
Pump 2 Software General Alarm
RPS_BSP2_PF_ZGA
RET
Return from Subroutine

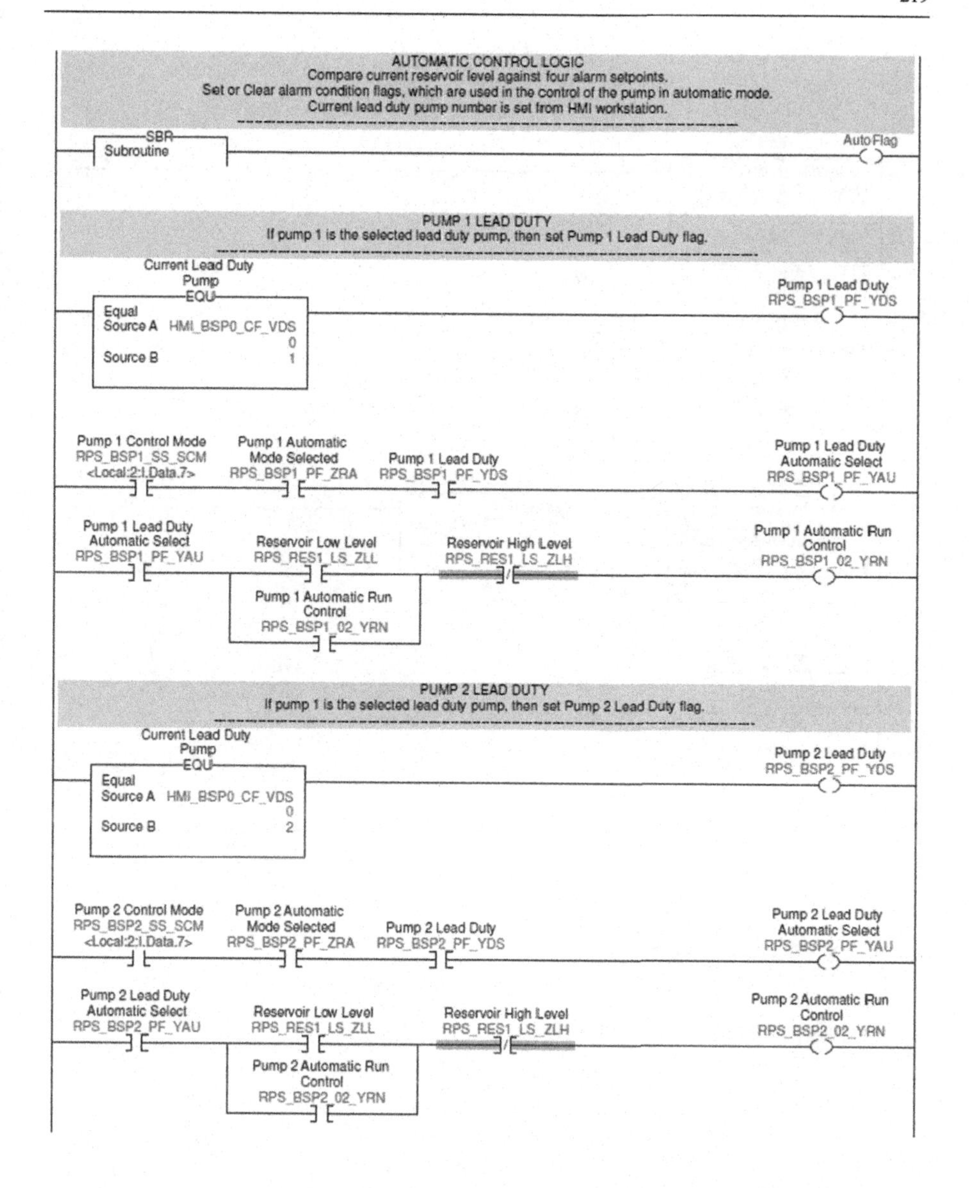

AUTOMATIC CONTROL LOGIC
Compare current reservoir level against four alarm setpoints.
Set or Clear alarm condition flags, which are used in the control of the pump in automatic mode.
Current lead duty pump number is set from HMI workstation.
SBR
Subroutine
AutoFlag
PUMP 1 LEAD DUTY
If pump 1 is the selected lead duty pump, then set Pump 1 Lead Duty flag.
Current Lead Duty Pump
EQU
Equal
Source A HMI_BSP0_CF_VDS
0
Source B 1
Pump 1 Lead Duty
RPS_BSP1_PF_YDS
Pump 1 Control Mode
RPS_BSP1_SS_SCM
<Local:2:I.Data.7>
Pump 1 Automatic Mode Selected
RPS_BSP1_PF_ZRA
Pump 1 Lead Duty
RPS_BSP1_PF_YDS
Pump 1 Lead Duty Automatic Select
RPS_BSP1_PF_YAU
Pump 1 Lead Duty Automatic Select
RPS_BSP1_PF_YAU
Reservoir Low Level
RPS_RES1_LS_ZLL
Reservoir High Level
RPS_RES1_LS_ZLH
Pump 1 Automatic Run Control
RPS_BSP1_02_YRN
Pump 1 Automatic Run Control
RPS_BSP1_02_YRN
PUMP 2 LEAD DUTY
If pump 1 is the selected lead duty pump, then set Pump 2 Lead Duty flag.
Current Lead Duty Pump
EQU
Equal
Source A HMI_BSP0_CF_VDS
0
Source B 2
Pump 2 Lead Duty
RPS_BSP2_PF_YDS
Pump 2 Control Mode
RPS_BSP2_SS_SCM
<Local:2:I.Data.7>
Pump 2 Automatic Mode Selected
RPS_BSP2_PF_ZRA
Pump 2 Lead Duty
RPS_BSP2_PF_YDS
Pump 2 Lead Duty Automatic Select
RPS_BSP2_PF_YAU
Pump 2 Lead Duty Automatic Select
RPS_BSP2_PF_YAU
Reservoir Low Level
RPS_RES1_LS_ZLL
Reservoir High Level
RPS_RES1_LS_ZLH
Pump 2 Automatic Run Control
RPS_BSP2_02_YRN
Pump 2 Automatic Run Control
RPS_BSP2_02_YRN

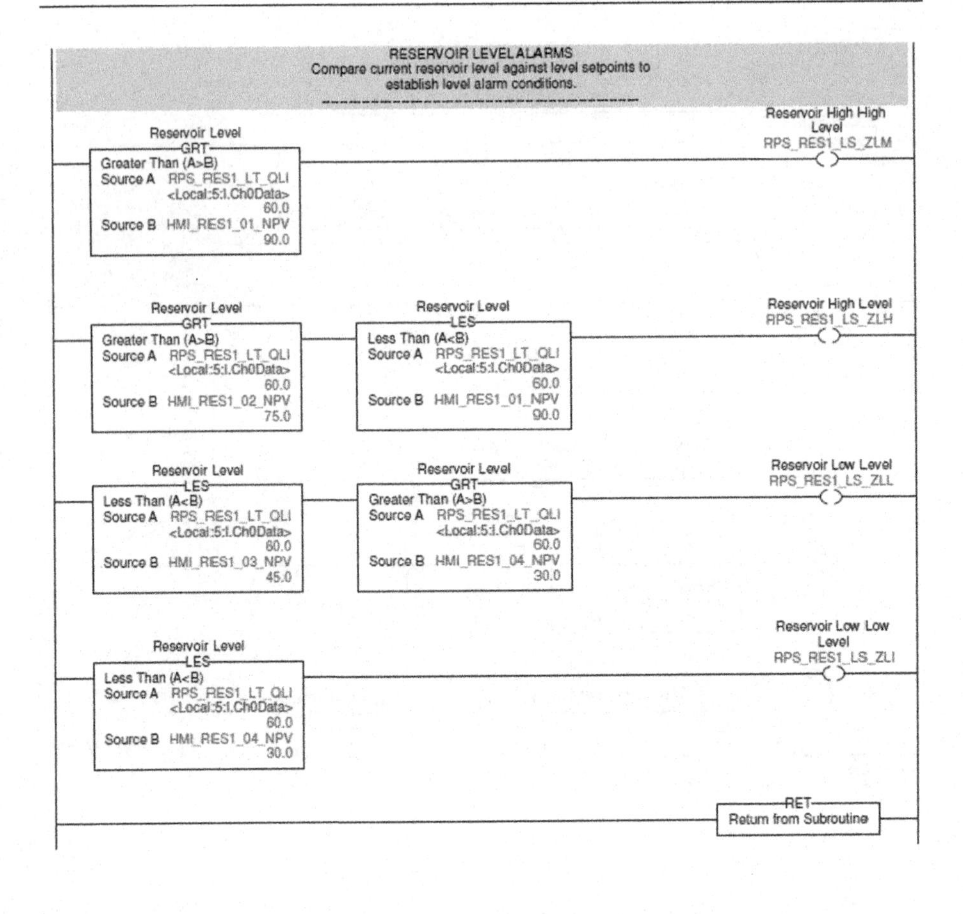

RESERVOIR LEVEL ALARMS
Compare current reservoir level against level setpoints to establish level alarm conditions.
Reservoir Level
GRT
Greater Than (A>B)
Source A RPS_RES1_LT_QLI
<Local:5:I.Ch0Data>
60.0
Source B HMI_RES1_01_NPV
90.0
Reservoir High High Level
RPS_RES1_LS_ZLM
Reservoir Level
GRT
Greater Than (A>B)
Source A RPS_RES1_LT_QLI
<Local:5:I.Ch0Data>
60.0
Source B HMI_RES1_02_NPV
75.0
Reservoir Level
LES
Less Than (A<B)
Source A RPS_RES1_LT_QLI
<Local:5:I.Ch0Data>
60.0
Source B HMI_RES1_01_NPV
90.0
Reservoir High Level
RPS_RES1_LS_ZLH
Reservoir Level
LES
Less Than (A<B)
Source A RPS_RES1_LT_QLI
<Local:5:I.Ch0Data>
60.0
Source B HMI_RES1_03_NPV
45.0
Reservoir Level
GRT
Greater Than (A>B)
Source A RPS_RES1_LT_QLI
<Local:5:I.Ch0Data>
60.0
Source B HMI_RES1_04_NPV
30.0
Reservoir Low Level
RPS_RES1_LS_ZLL
Reservoir Level
LES
Less Than (A<B)
Source A RPS_RES1_LT_QLI
<Local:5:I.Ch0Data>
60.0
Source B HMI_RES1_04_NPV
30.0
Reservoir Low Low Level
RPS_RES1_LS_ZLI
RET
Return from Subroutine

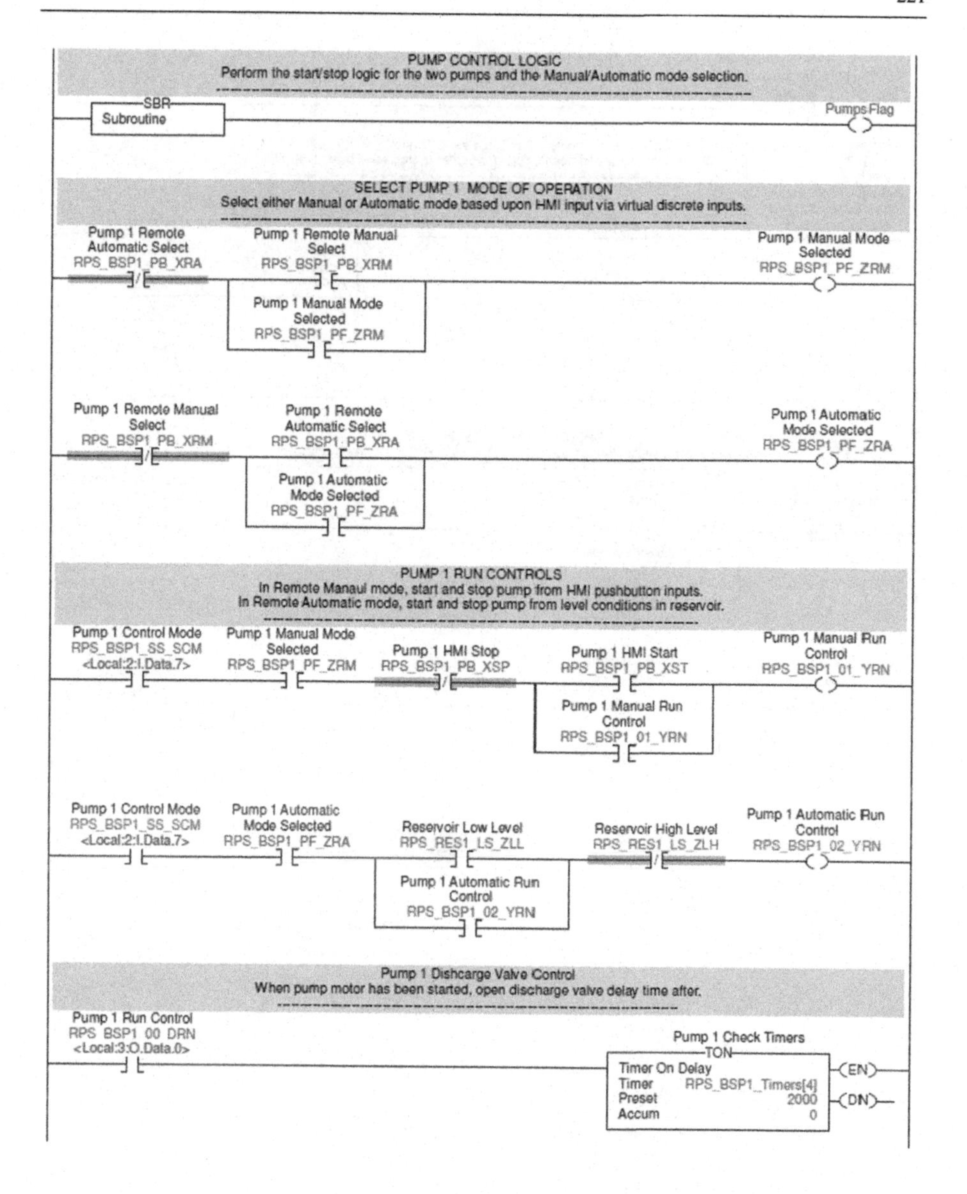
PUMP CONTROL LOGIC
Perform the start/stop logic for the two pumps and the Manual/Automatic mode selection.
SBR
Subroutine
Pumps Flag
SELECT PUMP 1 MODE OF OPERATION
Select either Manual or Automatic mode based upon HMI input via virtual discrete inputs.
Pump 1 Remote Automatic Select
RPS_BSP1_PB_XRA
Pump 1 Remote Manual Select
RPS_BSP1_PB_XRM
Pump 1 Manual Mode Selected
RPS_BSP1_PF_ZRM
Pump 1 Remote Manual Select
RPS_BSP1_PB_XRM
Pump 1 Remote Automatic Select
RPS_BSP1_PB_XRA
Pump 1 Automatic Mode Selected
RPS_BSP1_PF_ZRA
PUMP 1 RUN CONTROLS
In Remote Manaul mode, start and stop pump from HMI pushbutton inputs.
In Remote Automatic mode, start and stop pump from level conditions in reservoir.
Pump 1 Control Mode
RPS_BSP1_SS_SCM
<Local:2:I.Data.7>
Pump 1 Manual Mode Selected
RPS_BSP1_PF_ZRM
Pump 1 HMI Stop
RPS_BSP1_PB_XSP
Pump 1 HMI Start
RPS_BSP1_PB_XST
Pump 1 Manual Run Control
RPS_BSP1_01_YRN
Pump 1 Automatic Mode Selected
RPS_BSP1_PF_ZRA
Reservoir Low Level
RPS_RES1_LS_ZLL
Reservoir High Level
RPS_RES1_LS_ZLH
Pump 1 Automatic Run Control
RPS_BSP1_02_YRN
Pump 1 Dishcarge Valve Control
When pump motor has been started, open discharge valve delay time after.
Pump 1 Run Control
RPS_BSP1_00_DRN
<Local:3:O.Data.0>
Pump 1 Check Timers
TON
Timer On Delay
Timer RPS_BSP1_Timers[4]
Preset 2000
Accum 0
EN
DN

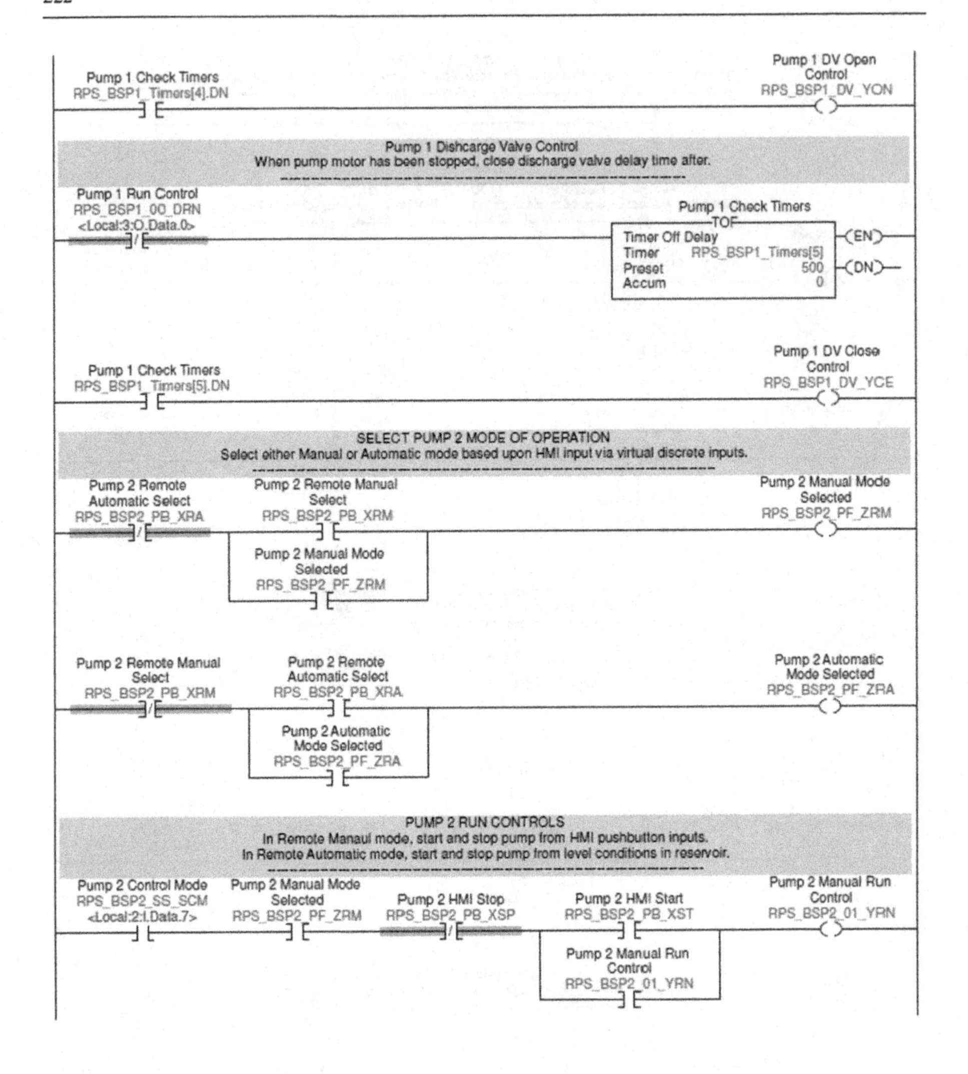
Pump 1 Check Timers
RPS_BSP1_Timers[4].DN
Pump 1 DV Open Control
RPS_BSP1_DV_YON
Pump 1 Dishcarge Valve Control
When pump motor has been stopped, close discharge valve delay time after.
Pump 1 Run Control
RPS_BSP1_00_DRN
<Local:3:O.Data.0>
Pump 1 Check Timers
TOF
Timer Off Delay
Timer RPS_BSP1_Timers[5]
Preset 500
Accum 0
EN
DN
Pump 1 Check Timers
RPS_BSP1_Timers[5].DN
Pump 1 DV Close Control
RPS_BSP1_DV_YCE
SELECT PUMP 2 MODE OF OPERATION
Select either Manual or Automatic mode based upon HMI input via virtual discrete inputs.
Pump 2 Remote Automatic Select
RPS_BSP2_PB_XRA
Pump 2 Remote Manual Select
RPS_BSP2_PB_XRM
Pump 2 Manual Mode Selected
RPS_BSP2_PF_ZRM
Pump 2 Manual Mode Selected
RPS_BSP2_PF_ZRM
Pump 2 Remote Manual Select
RPS_BSP2_PB_XRM
Pump 2 Remote Automatic Select
RPS_BSP2_PB_XRA
Pump 2 Automatic Mode Selected
RPS_BSP2_PF_ZRA
Pump 2 Automatic Mode Selected
RPS_BSP2_PF_ZRA
PUMP 2 RUN CONTROLS
In Remote Manaul mode, start and stop pump from HMI pushbutton inputs.
In Remote Automatic mode, start and stop pump from level conditions in reservoir.
Pump 2 Control Mode
RPS_BSP2_SS_SCM
<Local:2:I.Data.7>
Pump 2 Manual Mode Selected
RPS_BSP2_PF_ZRM
Pump 2 HMI Stop
RPS_BSP2_PB_XSP
Pump 2 HMI Start
RPS_BSP2_PB_XST
Pump 2 Manual Run Control
RPS_BSP2_01_YRN
Pump 2 Manual Run Control
RPS_BSP2_01_YRN

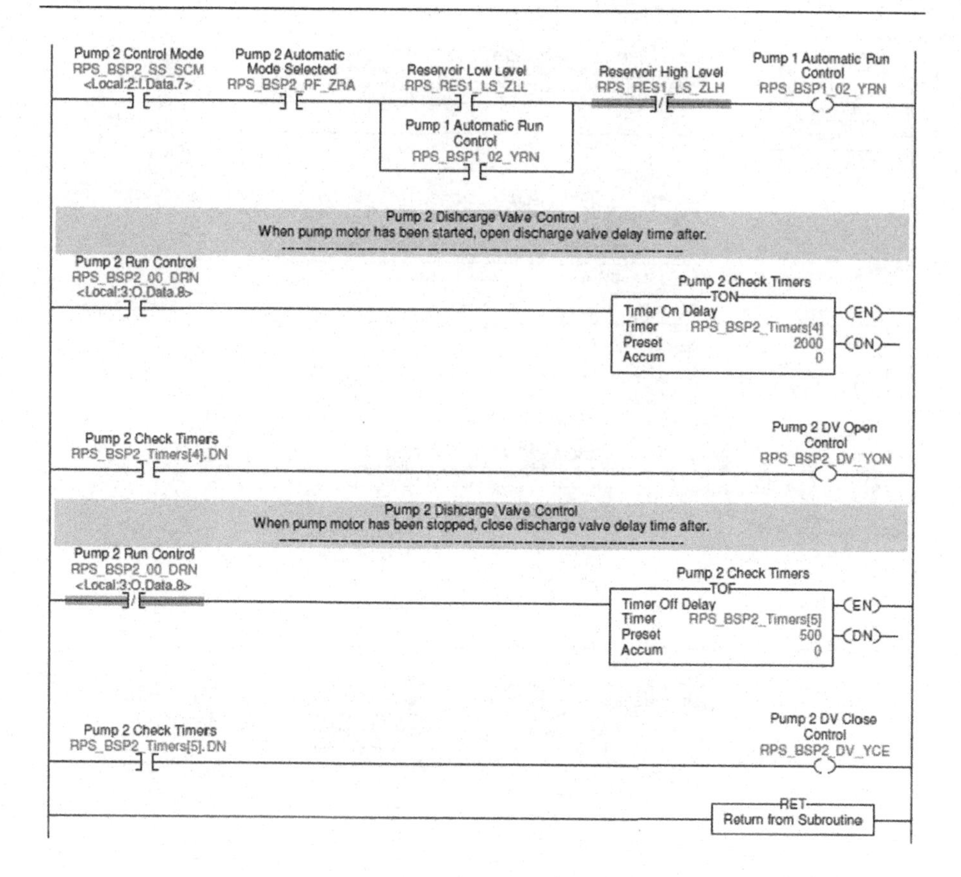
Pump 2 Control Mode
RPS_BSP2_SS_SCM
<Local:2:I.Data.7>
Pump 2 Automatic Mode Selected
RPS_BSP2_PF_ZRA
Reservoir Low Level
RPS_RES1_LS_ZLL
Pump 1 Automatic Run Control
RPS_BSP1_02_YRN
Reservoir High Level
RPS_RES1_LS_ZLH
Pump 1 Automatic Run Control
RPS_BSP1_02_YRN
Pump 2 Dishcarge Valve Control
When pump motor has been started, open discharge valve delay time after.
Pump 2 Run Control
RPS_BSP2_00_DRN
<Local:3:O.Data.8>
Pump 2 Check Timers
TON
Timer On Delay
Timer RPS_BSP2_Timers[4]
Preset 2000
Accum 0
EN
DN
Pump 2 Check Timers
RPS_BSP2_Timers[4].DN
Pump 2 DV Open Control
RPS_BSP2_DV_YON
Pump 2 Dishcarge Valve Control
When pump motor has been stopped, close discharge valve delay time after.
Pump 2 Run Control
RPS_BSP2_00_DRN
<Local:3:O.Data.8>
Pump 2 Check Timers
TOF
Timer Off Delay
Timer RPS_BSP2_Timers[5]
Preset 500
Accum 0
EN
DN
Pump 2 Check Timers
RPS_BSP2_Timers[5].DN
Pump 2 DV Close Control
RPS_BSP2_DV_YCE
RET
Return from Subroutine

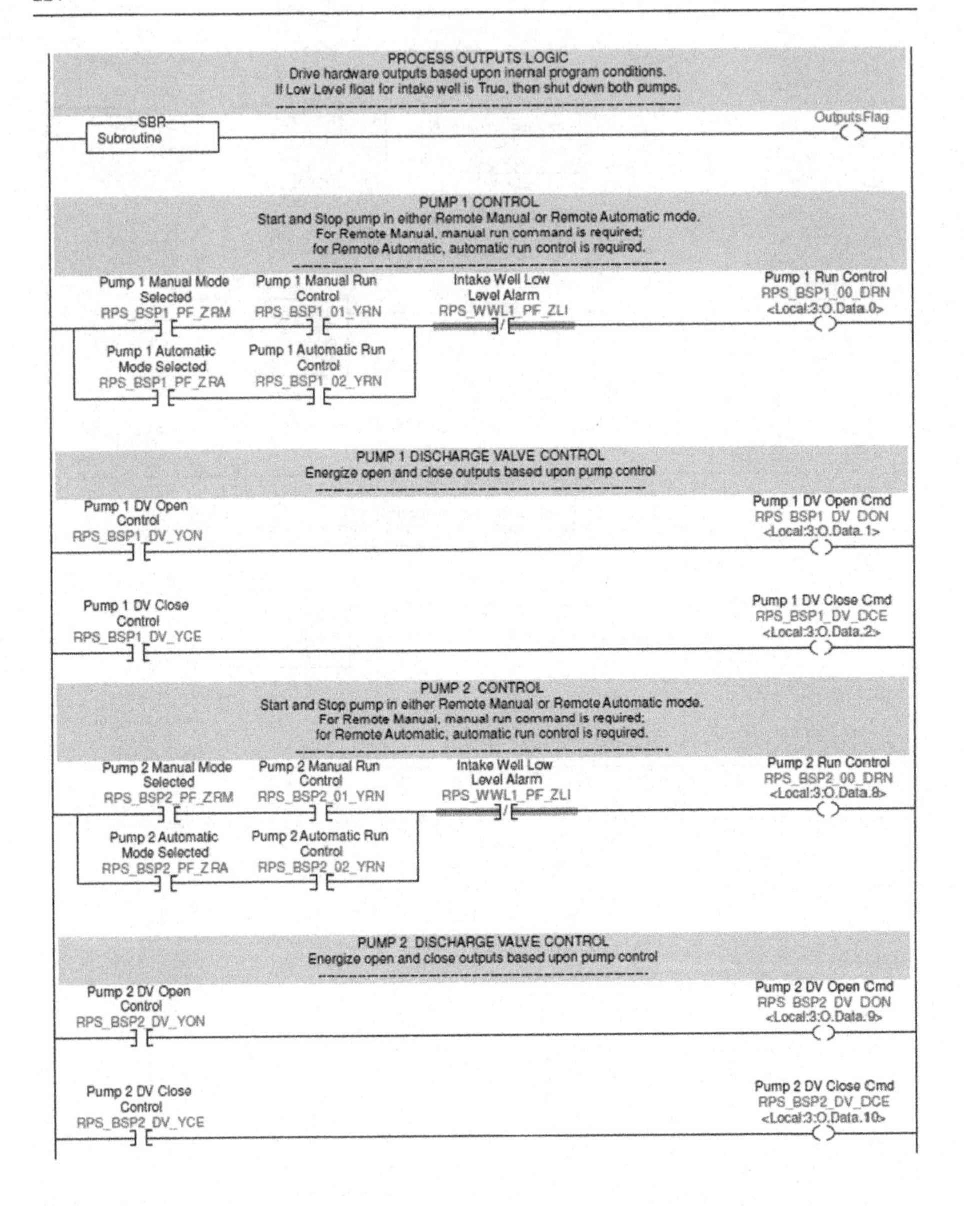

PROCESS OUTPUTS LOGIC
Drive hardware outputs based upon inernal program conditions.
If Low Level float for intake well is True, then shut down both pumps.
SBR
Subroutine
OutputsFlag
PUMP 1 CONTROL
Start and Stop pump in either Remote Manual or Remote Automatic mode.
For Remote Manual, manual run command is required;
for Remote Automatic, automatic run control is required.
Pump 1 Manual Mode Selected
RPS_BSP1_PF_ZRM
Pump 1 Manual Run Control
RPS_BSP1_01_YRN
Intake Well Low Level Alarm
RPS_WWL1_PF_ZLI
Pump 1 Run Control
RPS_BSP1_00_DRN
<Local:3:O.Data.0>
Pump 1 Automatic Mode Selected
RPS_BSP1_PF_ZRA
Pump 1 Automatic Run Control
RPS_BSP1_02_YRN
PUMP 1 DISCHARGE VALVE CONTROL
Energize open and close outputs based upon pump control
Pump 1 DV Open Control
RPS_BSP1_DV_YON
Pump 1 DV Open Cmd
RPS_BSP1_DV_DON
<Local:3:O.Data.1>
Pump 1 DV Close Control
RPS_BSP1_DV_YCE
Pump 1 DV Close Cmd
RPS_BSP1_DV_DCE
<Local:3:O.Data.2>
PUMP 2 CONTROL
Start and Stop pump in either Remote Manual or Remote Automatic mode.
For Remote Manual, manual run command is required;
for Remote Automatic, automatic run control is required.
Pump 2 Manual Mode Selected
RPS_BSP2_PF_ZRM
Pump 2 Manual Run Control
RPS_BSP2_01_YRN
Intake Well Low Level Alarm
RPS_WWL1_PF_ZLI
Pump 2 Run Control
RPS_BSP2_00_DRN
<Local:3:O.Data.8>
Pump 2 Automatic Mode Selected
RPS_BSP2_PF_ZRA
Pump 2 Automatic Run Control
RPS_BSP2_02_YRN
PUMP 2 DISCHARGE VALVE CONTROL
Energize open and close outputs based upon pump control
Pump 2 DV Open Control
RPS_BSP2_DV_YON
Pump 2 DV Open Cmd
RPS_BSP2_DV_DON
<Local:3:O.Data.9>
Pump 2 DV Close Control
RPS_BSP2_DV_YCE
Pump 2 DV Close Cmd
RPS_BSP2_DV_DCE
<Local:3:O.Data.10>

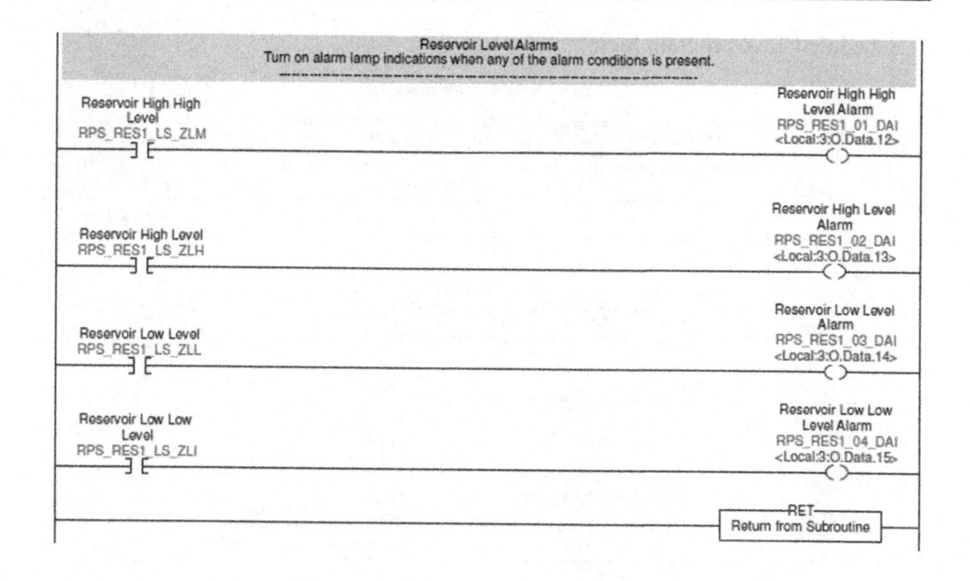

Reservoir Level Alarms
Turn on alarm lamp indications when any of the alarm conditions is present.
Reservoir High High Level
RPS_RES1_LS_ZLM
Reservoir High High Level Alarm
RPS_RES1_01_DAI
<Local:3:O.Data.12>
Reservoir High Level
RPS_RES1_LS_ZLH
Reservoir High Level Alarm
RPS_RES1_02_DAI
<Local:3:O.Data.13>
Reservoir Low Level
RPS_RES1_LS_ZLL
Reservoir Low Level Alarm
RPS_RES1_03_DAI
<Local:3:O.Data.14>
Reservoir Low Low Level
RPS_RES1_LS_ZLI
Reservoir Low Low Level Alarm
RPS_RES1_04_DAI
<Local:3:O.Data.15>
RET
Return from Subroutine

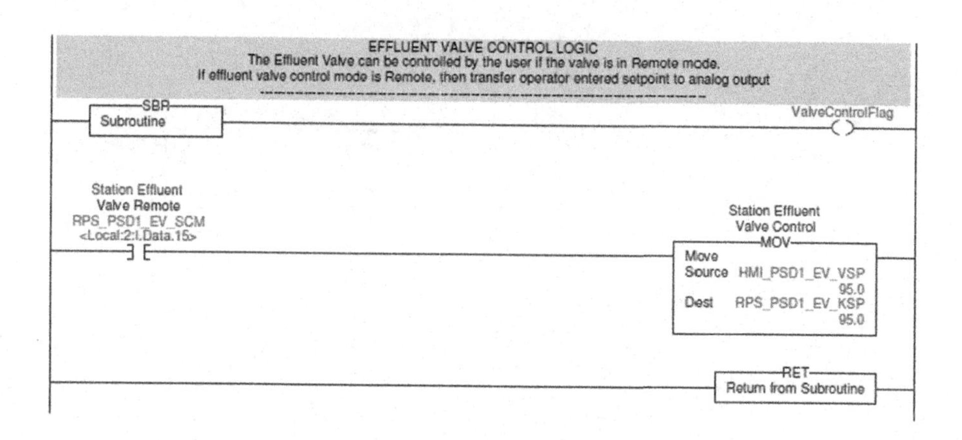

EFFLUENT VALVE CONTROL LOGIC
The Effluent Valve can be controlled by the user if the valve is in Remote mode.
If effluent valve control mode is Remote, then transfer operator entered setpoint to analog output
SBR
Subroutine
ValveControlFlag
Station Effluent Valve Remote
RPS_PSD1_EV_SCM
<Local:2:I.Data.15>
Station Effluent Valve Control
MOV
Move
Source HMI_PSD1_EV_VSP
95.0
Dest RPS_PSD1_EV_KSP
95.0
RET
Return from Subroutine

Scheduled Task program logic:

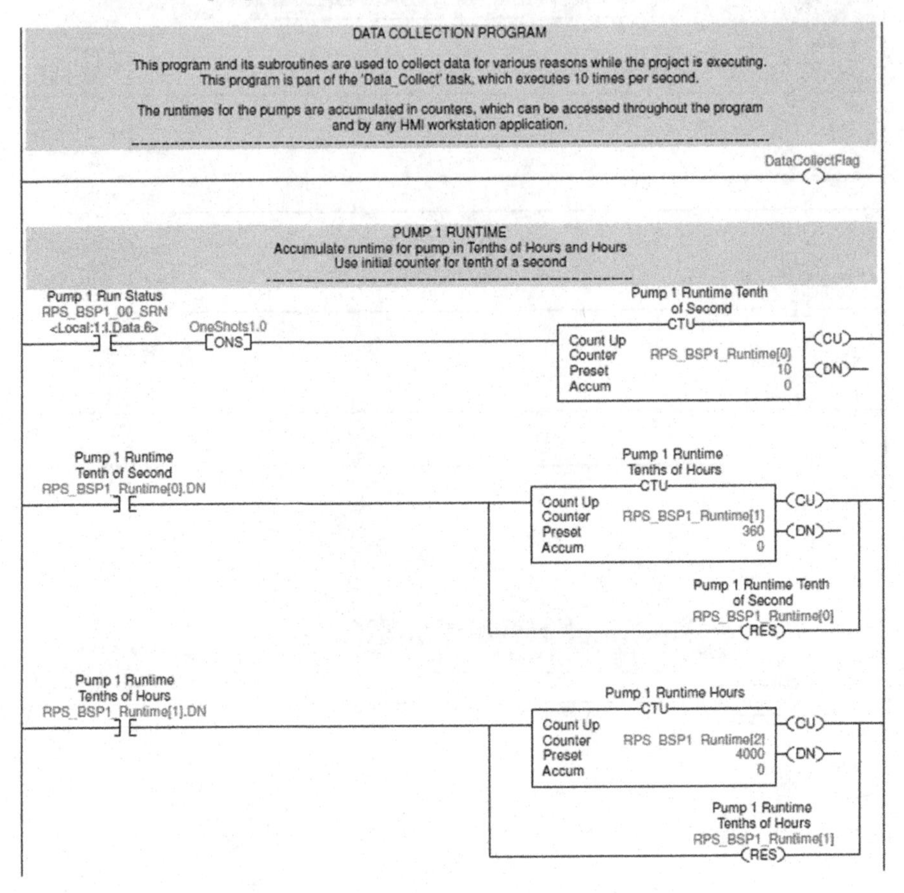
DATA COLLECTION PROGRAM
This program and its subroutines are used to collect data for various reasons while the project is executing.
This program is part of the 'Data_Collect' task, which executes 10 times per second.
The runtimes for the pumps are accumulated in counters, which can be accessed throughout the program and by any HMI workstation application.
DataCollectFlag
PUMP 1 RUNTIME
Accumulate runtime for pump in Tenths of Hours and Hours
Use initial counter for tenth of a second
Pump 1 Run Status
RPS_BSP1_00_SRN
<Local:1:I.Data.6>
OneShots1.0
ONS
Pump 1 Runtime Tenth of Second
CTU
Count Up
Counter RPS_BSP1_Runtime[0]
Preset 10
Accum 0
CU
DN
Pump 1 Runtime Tenth of Second
RPS_BSP1_Runtime[0].DN
Pump 1 Runtime Tenths of Hours
CTU
Count Up
Counter RPS_BSP1_Runtime[1]
Preset 360
Accum 0
CU
DN
Pump 1 Runtime Tenth of Second
RPS_BSP1_Runtime[0]
RES
Pump 1 Runtime Tenths of Hours
RPS_BSP1_Runtime[1].DN
Pump 1 Runtime Hours
CTU
Count Up
Counter RPS_BSP1_Runtime[2]
Preset 4000
Accum 0
CU
DN
Pump 1 Runtime Tenths of Hours
RPS_BSP1_Runtime[1]
RES

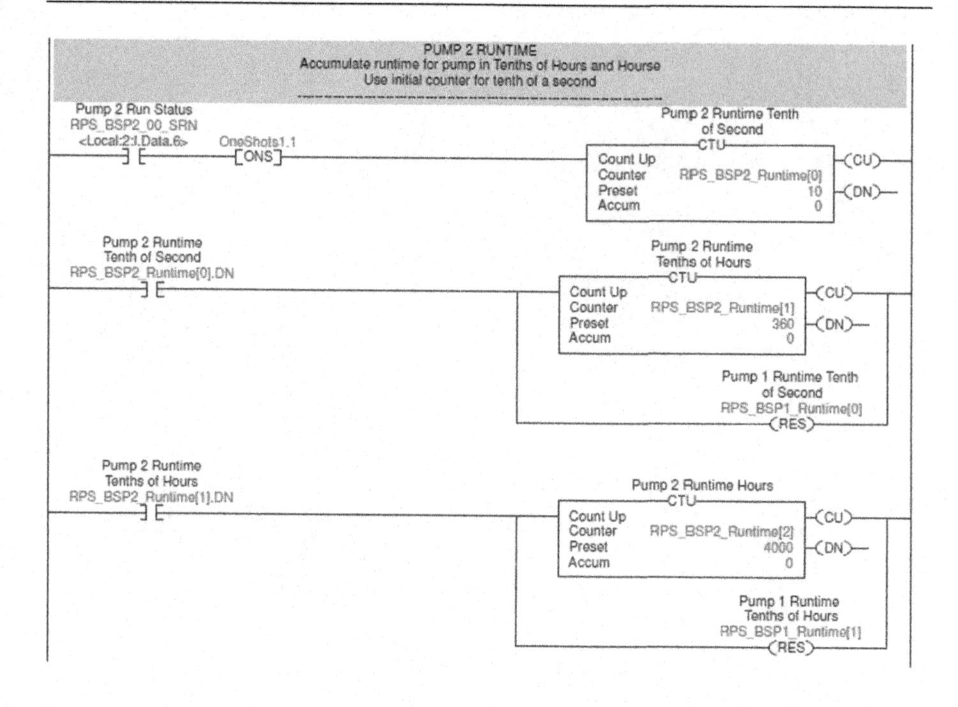

PUMP 2 RUNTIME
Accumulate runtime for pump in Tenths of Hours and Hourse
Use initial counter for tenth of a second
Pump 2 Run Status
RPS_BSP2_00_SRN
<Local:2:I.Data.6>
OneShots1.1
ONS
Pump 2 Runtime Tenth of Second
CTU
Count Up
Counter RPS_BSP2_Runtime[0]
Preset 10
Accum 0
CU
DN
Pump 2 Runtime Tenth of Second
RPS_BSP2_Runtime[0].DN
Pump 2 Runtime Tenths of Hours
CTU
Count Up
Counter RPS_BSP2_Runtime[1]
Preset 360
Accum 0
CU
DN
Pump 1 Runtime Tenth of Second
RPS_BSP1_Runtime[0]
RES
Pump 2 Runtime Tenths of Hours
RPS_BSP2_Runtime[1].DN
Pump 2 Runtime Hours
CTU
Count Up
Counter RPS_BSP2_Runtime[2]
Preset 4000
Accum 0
CU
DN
Pump 1 Runtime Tenths of Hours
RPS_BSP1_Runtime[1]
RES

Printed in the United States
By Bookmasters